D0197704

"The best book that I have read about grass-roots film marketing. Troma Entertainment actually uses the marketing secrets described in this excellent book, and has for nearly 30 years! They work!"

— Lloyd Kaufman
President, Troma Entertainment
Creator, *The Toxic Avenger*

"For those filmmakers in need of an audience and distribution deal, this book is a must-read. Thank God someone finally wrote it, and wrote it well. Mark Bosko just saved me, and those of us working in the industry, a great deal of time and effort."

— Charlie Humphrey
Executive Director, Pittsburgh Filmmakers

"Finding distribution is one of the toughest obstacles facing an independent filmmaker. This book is a must-read for anyone trying to overcome that barrier to success. Bosko also does a great job giving real-life examples to instruct the reader how to make a profit with their independent feature or video."

— Eric Colley
President and Owner, IndieClub

"Very well written and full of common-sense advice that helps illuminate the often confounding independent film distribution process."

— Nancy Gerstman
Co-President, Zeitgeist Films Ltd.

"Mark Bosko's book defines what every filmmaker should study and learn even before undertaking the initial task of production. Making a movie is the easy part – distribution, marketing, and sales are what we all work toward. This book details invaluable procedures that every serious film and videomaker cannot do without."

— Ted V. Mikels
Legendary Independent Filmmaker
Writer-Producer-Director of
The Corpse Grinders, The Doll Squad,
and *The Black Klansman*

"Mark Bosko's book is extremely well researched and replete with actual case histories and insightful observations and advice. If you are producing and marketing an independent film, Bosko's book should be on your must-read list."

— Harris Tulchin, Harris Tulchin and Associates
Entertainment Attorney, Producer's Rep, Author
www.medialawyer.com

"Get this book if you have a finished film you want to sell!!! It's packed full with incredible ideas, gimmicks, and ploys to get little films out there and noticed. I wish we had this when we were doing *Blair Witch*."

— Eduardo Sanchez
Co-Writer, Co-Director, Co-Editor of
The Blair Witch Project

"Many independent filmmakers think that if they make a great film, the world will come to their doorstep. This is just not so. Bosko's definitive book on film marketing shows you in great detail how to take your film into the world. Read this book BEFORE making your film so you can enhance the creative aspects of the film to help it reach an audience. If you are a serious filmmaker and want your work to be seen, this is a must-have book."

— Michael Wiese
Filmmaker, Author of
Film and Video Marketing

"Not only independent filmmakers, but screenwriters, too, will find this book one heck of a terrific bible on marketing."

— Sable Jak, *Scr(i)pt Magazine*

"Clear, comprehensive, and written in an informal, relaxed style, this book is a good first step for anyone even thinking of getting into the indie film business."

— Cynthia Close
Executive Director
Documentary Educational Resources

"To achieve true success in today's shark-infested waters of the world entertainment industry, read and follow the concise and no-nonsense guidelines and vital insider tips in Mark Bosko's *The Complete Independent Movie Marketing Handbook*."

— Elizabeth English
Founder & Executive Director
Moondance International Film Festival

MICHAEL WIESE PRODUCTIONS
www.mwp.com

Since 1981, Michael Wiese Productions has been dedicated to providing novice and seasoned filmmakers with vital information on all aspects of filmmaking and videomaking. We have published more than 60 books, used in over 500 film schools worldwide.

Our authors are successful industry professionals — they believe that the more knowledge and experience they share with others, the more high-quality films will be made. That's why they spend countless hours writing about the hard stuff: budgeting, financing, directing, marketing, and distribution. Many of our authors, including myself, are often invited to conduct filmmaking seminars around the world.

We truly hope that our publications, seminars, and consulting services will empower you to create enduring films that will last for generations to come.

We're here to help. Let us hear from you.

Sincerely,

Michael Wiese
Publisher, Filmmaker

THE COMPLETE
INDEPENDENT
MOVIE
MARKETING
HANDBOOK

PROMOTE, DISTRIBUTE & SELL YOUR FILM OR VIDEO

MARK STEVEN BOSKO

Published by Michael Wiese Productions
11288 Ventura Blvd., Suite 621
Studio City, CA 91604
tel. (818) 379-8799
fax (818) 986-3408
mw@mwp.com
www.mwp.com

Cover Design: AG Design Company
Book Layout: Gina Mansfield
Editor: Brett Jay Markel

Printed by McNaughton & Gunn, Inc., Saline, Michigan
Manufactured in the United States of America

© 2003 Mark Steven Bosko

All rights reserved. No part of this book may be reproduced in any form or by any means without permission in writing from the publisher, except for the inclusion of brief quotations in a review.

ISBN # 0-941188-76-0

Library of Congress Cataloging-in-Publication Data

Bosko, Mark Steven, 1965-
    The complete independent movie marketing handbook : promote,
distribute, and sell your film or video / Mark Steven Bosko.
        p. cm.
    ISBN 0-941188-76-0
    1. Motion pictures–Marketing–Handbooks, manuals, etc. 2. Video
recordings–Marketing–Handbooks, manuals, etc. I. Title.
    PN1995.9.M29 B67 2003
    384'.8'0688–dc21
                                        2002154125

DEDICATION

For Tawnya — my continued inspiration

ACKNOWLEDGEMENTS

The author would like to graciously thank the following individuals and companies for their assistance with and contributions toward the production of this book: Dana Archer, Berlin Productions, Arthur Borman, Kat Candler, Ted Chalmers, Cinema Apocalypse, Eric Colley, T. Michael Conway, Creative Light Entertainment, Creative Light Worldwide, Darlene Cysper, *DirectorsCut.com*, John Ervin, Highland Myst Films, *IndieClub.com*, IndieWire, Inferno Film Productions, John P. Katsantonis, The Katsantonis Group, John Keeyes, David Larry, Mean Time Productions, Michael Mongillo, Mutiny Productions, Roger Nygard, Jimi Petulla, Craig James Photography, and Todd Wardrope. Additionally, the author is grateful for the faithful support of his parents, Rose and Francisco Bosko.

FOREWORD

BY CHRIS GORE

When you've been around the independent film scene as long as I have, you're often hit up to do a lot of things for free. Things like acting in movies, working on film sets, speaking on panels, or writing the foreword for a book called *The Complete Independent Movie Marketing Handbook*. Trouble is, I don't like to do anything for free unless I feel very strongly about it. When Mark Bosko approached me about this book that you're holding (well, hopefully, you're not just holding it, but you've actually purchased it), he was merely seeking some sort of positive quote that he might use to promote it. I told him that I'd take a look.

After absorbing Mark's no-nonsense approach to the very complex subject of movie marketing, I told him that not only would I tell every filmmaker I knew that they simply must read his book, but that I wanted to write the foreword — in fact, the one that you are reading right now. But, let's face it, most people don't actually *read* the forewords of books. I realize that only a handful of people who've actually purchased this book are even bothering to read the words I am typing at this very moment. So, if you are one of the lucky few engrossed in this essay, I can let you in on a few secrets. Okay, one very large secret. Indie filmmakers need this book, and there's a very important reason.

This reason is so important, I implore you to write it down. Or don't write it down — either remember it, or highlight this or something, but the next thing I have to say is critically important and will make reading this foreword all the more worthwhile: Becoming a successful filmmaker is a monumental task. The reason is that it not only requires you to make a creative triumph on screen (and the odds are stacked against you doing this already), you must also be very business-minded to achieve true success.

What is true success? For some, it can be getting the first movie made and distributed. For others, it can be just getting a second project off the ground. To me, true success is becoming a working filmmaker who makes a living making movies. That's right, getting paid to make films. And there are very few in the indie film world who achieve this kind of success with regularity. It requires one to be both an incredibly creative person *and* a savvy businessperson. The troubling reality is that most filmmakers tend to focus too much on either one or the other. Focus more on the *creative*, and you may really have something wonderful; but without the business skills, you will run into trouble. If you put all your energy into the *business*, you may have a balanced checkbook, but a film which does not wow anyone. The truly successful filmmakers strike a balance between being both creative-minded and business-minded. Remember that. This book is the first step in your journey toward reaching that balance and that success.

I'll be looking for all your well-marketed films at the theater or the next festival, or on cable or DVD. And you better have listened to everything Mark has said or you just wasted your cash. Don't just read this book, its ideas and concepts should be burned into your memory.

Oh, yeah, the publisher wanted to be sure that I mentioned things like what makes this book valuable and unique and why indie filmmakers simply must purchase it. Put simply, this book will teach you nothing about how to make a film. But it will teach you everything you need to know to get your film out there. In many ways, it's almost more important than making the film itself. Mark has done a service to the indie film community for taking the enormous amount of time and energy it takes to put a project like this together. (Believe me, I know.) But don't thank me. Thank Mark. He wrote the damn thing. I just wrote this foreword at the last minute.

Let me throw out just one last piece of advice: No matter the obstacles, never quit.

Chris Gore is the editor of *FilmThreat.com* and author of *The Ultimate Film Festival Survival Guide*. Click on www.filmthreat.com

P.S. I like to think of this book as, in some way, a companion to *The Ultimate Film Festival Survival Guide*. I'm grateful to Mark for writing this book, so I don't have to. By reading both, indie filmmakers will be equipped with all the business knowledge they need to find success in the indie film world. (I was also seeking a way to throw in another plug for my book; I know this P.S. is not very creative, but it's the best I can do under the strict deadline I was given to write this foreword.)

P.S.S. One last thing, not for you, this is for the author: Hey Mark, you owe me a beer.

INTRODUCTION

Without distribution, you might as well leave the lens cap on. Harsh? Not really. Unless, of course, you are a video hobbyist or film "artiste" with no ambitions of working in the entertainment industry. And there's nothing wrong with that, except you've picked up the wrong book. You see, the only way to continue making films and videos is to profit (both monetarily and career-wise) from your films and videos. That will never happen unless you have the means to find an audience willing to pay for what you've created. This is what's known as **distribution**.

With experience as both a filmmaker and videomaker, there are three things I've learned about distribution:

1. You can't sell a film or video without distribution.
2. You can't distribute a film or video without promotion.
3. And you can't promote a film or video without some damn hard work.

In other words, distribution isn't easy. It's not for the light-hearted, the lazy, or the uninspired. But then again, neither is the act of putting together an independent film or video. You have to be every bit as committed to getting your project in front of an audience as you were in getting it made. Unless you have this sort of fanaticism with entering or continuing in your position as a professional full-time film/video-maker, you probably wouldn't be reading this and every other text on the subject that comes along. You want to learn everything possible that will contribute to your success in the industry; hopefully, this book will help do just that.

The information on the following pages, however, is going to be different than what you may have gleaned from other books. To begin with, I'll assume you fall into one of the following situations:

1. You already have a finished production, or are well on your way to finishing one.
2. You have funds available and/or the equipment needed to produce your film or video.

3. If neither of the above, you have laid the preproduction
   groundwork necessary for the film/videomaking process.

With that in mind, this book will forego any discussions on production,
direction, editing, and so on. Nor will it discuss methods of attracting
investors and financing, recruiting talent, choosing film/tape formats,
or creating a production budget. All of these are subjects deserving
entire texts dedicated just to their study, and many fine choices are
available on these topics through booksellers and libraries everywhere.
This book deals strictly with the promotion, distribution, and sales of
your independent films and videos, offering step-by-step instruction to
techniques that you can truly accomplish.

Production is a complicated and tiring enough process in itself, and if
you have finished a film or video, then you should be congratulated on
this Herculean effort. Many people "talk" a good project, but it takes
a whole different measure of person to actually see it through to the
end. Be proud of yourself. For those of you educating yourselves, planning a production, or in the middle of shooting, just getting started is a
sign of your commitment. Taking that first step is often the hardest
part. Whether you are busy finishing your independent feature, documentary, or instructional video, or have a completed project "in the
can," you probably came to realize very quickly that time and money are
short. You want to get your hands on useful knowledge in the fastest
means possible, and you don't want it buried under pages of anecdotes
from billionaire Hollywood producers; idealistic, non-working "professionals in the field;" or technology-obsessed, out-of-touch videomakers.
You need to know how to get your project into the hands of distributors,
the media, video buyers, audiences, and others — now.

This book details street-level marketing techniques that can be used
by all film and videomaking professionals, regardless of your level of
experience or stage in the production process. The book contains
information in three settings:

1. Instructional — teaching you to look at your film or video as a
   product, as well as how to promote, distribute, and sell it

2. Practical – giving insight into the state of the independent film and video world through industry data and real-world examples
3. Referential – providing details on distributors, media, duplicators, and other sources indispensable to your success in the field

The workbook-like style and straightforward approach gives you the information you need to begin successfully marketing your independent film or video as soon as you begin reading. By following through the text, you'll learn how to promote your feature or video project, where to sell it, what price to ask, when to seek help, and when to quit... never! Let's get started:

INDEPENDENT FILMS AND VIDEOS

Just what is an independent film or video? If you go to Hollywood, they'll tell you an "independent" film is anything under $10 million. And on Madison Avenue, "independent" instructional tapes are churned out to the tune of $5,000 per finished minute. If only we were so lucky. To ground ourselves in reality and for purposes of this book, we will define an independent film or video as one with the following characteristics:

1. An operating budget of $1 million or less. Less being the key. Many of you will be working with budgets ranging from $500 to $50,000, and that's fine!
2. No stars or name actors to speak of. (This doesn't include genre-related "B," "C," or "Z" list talent)
3. Mainly non-studio, location shoots with small crews of 5 persons or less.
4. Often genre or exploitative themes such as horror, teen-comedy, popular fads, or current events.
5. Produced on various video, 8mm, or 16mm film formats.

The main focus of this book is promoting, distributing, and selling independent feature films — of which, according to *IndieWire*, an authority on the independent filmmaking community, more than 3,000 are made every year. However, the techniques described herein can also be effectively used with documentaries, instructional tapes, and any other filmed or videotaped productions intended for mass consumption that meet the numbered criteria above. Though these kinds of projects' inherent form is different from that of a feature-length film (shorter running length and an absence of fictional storyline are two immediate differences), productions such as these can be promoted, distributed, and sold in much the same manner, using many of the same channels as an independent movie.

THE VALUE OF INDEPENDENT FILMS AND VIDEOS

While most truly independent films and videos bear the disdain of "serious" critics and many lofty-minded industry individuals willing to only associate with (or at least admit to associating with) "A" projects, so-called "indies" continue to provide quality entertainment and instruction, and have, over the years, contributed significantly to American film history.

Where would Francis Ford Coppola be today if he didn't cut his teeth on the $29,000 *Dementia 13*? Do you think we would know Steven Spielberg if not for his training with TV films such as *Duel*? And, David Fincher, the director of *Se7en*, *Fight Club*, and *Alien³* got started in the business by shooting cheap music videos for New York City rappers. Need we even mention Daniel Myrick and Eduardo Sanchez, the guys behind the $60,000, shot-on-video *Blair Witch Project*? Quite noble beginnings for some of the biggest players in the entertainment industry.

If you delve into the background of most every successful film and videomaker working today, you'll certainly find an independent film or video production that was the beginning of it all. And if you think about it, that makes sense. Just as every other accomplished professional must gain competence through training and practice, so must the burgeoning entertainment producer. Don't ever apologize for the limitations — budget or otherwise — of your project. Think of every experience as a stepping-stone to your eventual acclaim and success in a tough industry. Sure, you may be working with a couple hundred bucks, semi-standard equipment, and homemade sets. But it is the effort that you put into the production that will eventually spell out its success or failure. You must believe that even though you are faced with what seem to be insurmountable hardships, your movie or video will emerge as a product of your tireless efforts to make a dream a reality.

That is basically what an independent film or video is all about anyhow — dreams. Sure, there are people churning out films and instructional tapes just for the money. It's a day job to these folks, not a passion. But I want to believe you have a different reason for working so hard. You want to create your "vision" and give it to the people of the world, so to speak. You want to feel the thrill of watching people react to something you made. You, like so many other producers, share the unequalled high attained from this experience. And that is why promotion, distribution, and sales are so vital to continued success.

Having produced the most effective exercise tape, emotionally moving motion picture, or socially conscious documentary means nothing if nobody knows it exists. You need an interested audience, a way to get it to those people, and someone to pay you for getting it there, or — to be brutally honest — you probably shouldn't have wasted your time producing the film or video in the first place. Don't get me wrong, artistic expression is great. But spending a year of your time on a project that is never shown beyond the confines of your own home isn't very satisfying.

That's why understanding and utilizing the many available promotional tools is so important. It is the first step in attracting attention to your "product." The first step in finding avenues to reach an audience. The first step in locating interested viewers. And the first step in getting paid for your labors (and you thought this was all just for fun).

GOOD TIMES, BAD TIMES...

Before we go any further, it's a good time to point out the demands and potential risks involved in the process of promoting, distributing, and selling your independent films and videos. Doing anything successfully takes time. You don't wake up as a major league baseball player. Nor do you become the lead singer for a Top 40 rock band after three voice lessons. Obviously, the same theory applies here. Your movies, instructional tapes, and documentaries are not going to find their way into rental stores, theatres, mass merchandisers, and on cable television with just a phone call or two. I only wish it were so simple. What was once "idle" time in your life will now be filled with phone calls, letter writing, and the general networking involved in attracting the right people to your project.

To most film and videomakers, this effort won't be an issue. Your so-called idle time is already consumed with production details, and this additional activity can be looked upon as another step in the process. However, unlike production, when filmed or taped images are a concrete testament to your many hours of labor, this phase often produces little to no results for quite some time. Basically, to those around you not intimately involved in what's going on, it will look like you are spinning your wheels.

Obviously, when you begin to promote your film, you'll be working your butt off. But that's not always evident, even to yourself. And this can lead to serious emotional strain. I'm not sure about you, but I'm not big on rejection. I try to avoid it at all costs. Let me correct that. I *tried* to avoid it at all costs. That was before I began to promote, distribute, and sell films and videos. You're gonna have to develop a thick skin — because in this business, rejection is unavoidable. Bottom line: You are going to be selling a subjective product, art even. And not everyone is going to be a fan. You can't let a setback bring you down, or you, and your effort, will suffer. Not to mention how your negative emotional state will affect those around you.

Along with the rejection comes inaction. I can't even begin to tell you how patient you're going to have to become to successfully pull this all off. While two weeks of waiting may seem like a lifetime to you, it's nothing to a distributor with a pile of tapes from potential suppliers sitting on his or her[1] desk.

One final thought: Keep a handle on expenses. Phone calls, postage, shipping supplies, screeners, and all of the other tools involved in promoting, distributing, and selling films and videos can add up fast. While the pages ahead include tons of methods to save money throughout this process, I haven't thought of everything. Just don't let your enthusiasm get the better of your common sense. If you're not sure what that means now, you will by the time you finish this book.

---

[1] This is the last time you'll read the phrase "his or her" or "he and she"; from this point on, I'll use one or the other pronoun and assume you understand that any sex could apply.

PROMOTING INDEPENDENT FILMS AND VIDEO

WHAT IS PROMOTION?

Promotion, and the role of a film promoter, is one of the oldest activities associated with the entertainment industry. Before multiplexes, home video, pay-per-view, and cable television, the financial success or failure of a film was due entirely to the efforts of its promoter.

Back in the good ol' days, feature films were carried (literally) from town to town by a film promoter. Viewings took place in revival tents, city hall buildings, and on outdoor screens with the lawn as a seating area. Projection equipment often broke down. Soundtracks, if present, were scratch-filled, tinny, and out-of-synch. Smoking was allowed. Admission might be a nickel. And, believe it or not, Sno-Caps® were not available in the concession area.

It definitely was a different experience than the stadium seating, $10 tickets, cappuccino- and nacho-filled, THX-powered, 70-foot screen movie palaces audiences are accustomed to today. But one thing hasn't changed: the role promotion plays in the "life" of a film.

In the dawn of what is now a multi-billion-dollar entertainment industry, advertising to the masses via national television networks, magazines, radio, and the Net wasn't available. Nor was the ability to "open" a movie simultaneously in 1,500 cities across the country. Still, promoting a film in those simpler times involved what it does now — the use of any and all methods and techniques available to interest potential viewers in the "product."

Whether that meant hiring "advance" men to poster a town days before the movie's premiere date, spreading sensationalistic (and usually false) word-of-mouth about the content of the film, or simply inviting key members of a community to a free screening (knowing they would influence the actions of other, ticket-buying audience members), a film promoter's role in selling seats could not be diminished. If the promoters didn't do their work, tickets wouldn't sell and the film would flop. If the film flopped, the promoter didn't get paid. That made it pretty simple for these folks to understand the importance of promotion as a part of the filmmaking process.

Times obviously have changed, and even though advance men, word-of-mouth, and free screenings still exist, many of the methods used to promote and market a film to potential audience members have changed as well. Today's motion picture studios have entire departments staffed with hundreds of marketing and promotions employees. At their access are million-dollar budgets to be spent on television advertising, in-theatre trailers, media junkets with the stars, promotional give-away items, Web sites, co-op contests, promotions with fast-food sponsors and the like — making it tough for anyone to go through a normal day without being exposed to at least one form of motion picture promotion.

For example, MGMs theatrical marketing department, which is responsible for the promotion of all the studios' product, consists of five functional groups: research, media planning, advertising, promotion, and publicity. The objective of the marketing department is to maximize each motion picture's commercial potential by designing and implementing a marketing campaign tailored to appeal to the movie's most receptive audience. The process begins before a film is completed, with the research department determining, through audience screenings and focus groups, a motion picture's appeal to its most likely target audience. The group also starts to develop marketing materials well in advance of a movie's scheduled theatrical release. A full-blown marketing campaign generally begins six months before release through the release of teaser trailers, posters, and in-theatre advertising materials. The campaign becomes more aggressive two to three months before release, as full-length trailers are distributed to theatres and additional materials are sent to theatre owners. Finally, a national media campaign is launched four to five weeks before opening day. This campaign generally involves advertising a picture's release on national television, including network prime time and syndication markets, national cable and radio, and in magazines, newspapers, and specific target markets. In addition, public appearances, such as television talk shows, are arranged for a picture's stars in order to promote the film. The entire process is managed by the in-house staff, although outside agencies are often used to provide creative services. Similar processes exist for MGMs release of home-video versions of motion pictures and ancillary products such as instructional tapes, syndicated television, and cable programming.

Now — if you were part of this kind of studio system — research, advertising, publicity, and all of the other tools associated with promoting a film or a video as described above, wouldn't be your concern. Regardless of whether the process succeeds or fails — if you are backed by a studio — you, as a filmmaker, still get paid. That's a great position to find yourself in.

But that's all a big "if." Most, if not all, of those people reading this book are independent film and videomakers lacking the support of a studio. At least at this point. You are more like the old-time film promoters with the success or failure of your individual efforts directly affecting both the amount of attention you garner and the money you make. You may get lucky and land a studio offer to distribute your independent movie. Or find a national cable channel begging for your video's broadcast rights. Right now, however, it's all up to you and how you use the information on the ensuing pages.

IS IT WORTH THE EFFORT?

The smart ones out there probably anticipated this concern already. Is all the work involved in promoting and marketing a film worth the effort? As the old gag goes, a film promoter is someone who gives the public what they want and then hopes they want it. Cynical? No doubt. But it really does seem like there's no guarantee for success.

Sure promotion is often more important to a motion picture than the film, or product, itself. Remember the monumentally hyped *Blair Witch Project*? Obviously it didn't live up to many audience members' expectations. But the promotional campaign attached to the movie made the public feel like they needed to see it. That's not always the case. Consider *Waterworld*, *Wild Wild West*, and *Town and Country*. These are examples of big-budgeted, star-studded films that failed to find an audience, even though millions of dollars were spent advertising and promoting them to what was believed to be a receptive and interested public.

Few people have ever bothered to try and quantify the reasons some films and videos fail and some succeed. While a number of factors that play into such an equation are known — cultural preferences, genre, the box office history of the leading actors, and whether the distribution is handled by a studio or an independent company — it appears mathematically impossible to come up with a formula to forecast the future of an unreleased movie or videotape based on its planned promotional campaign. Making the "Is it worth it?" question a tough one to answer. So studios keep on spending in the hopes that their actions will produce the next blockbuster. Besides, what's the alternative?

WHAT *IS* THE ALTERNATIVE?

Independent films and videos are a product, and like any other product, they need to be marketed. It is that simple. Why did you start using a Swiffer® instead of a broom? Why do you drink Dasani® water instead of the free stuff that comes out of your kitchen faucet? Why do you watch movies on DVD instead of VHS? All of these products and thousand of others that succeed in finding their "audience" do so because of successful marketing and promotional campaigns. And it's no different for films and videos. Entertainment products must find and satisfy an audience.

I can hear the artists stomping their feet already. Films and videos are products? And, worse yet, why worry about what the audience wants when you are creating art, right? Yes, art is good, but you need to be able to identify the commercial aspects of your project. You need to find those qualities that make it attractive to viewers, giving you something to sell it by. At this point in your life as a professional film or videomaker, the artistic aspects of your project might have to take a backseat (or, at least, a passenger seat) to the promotional aspects. You're going to have to dig past all the high-brow existentialistic meanings that you've attached to every frame of your film and ask why anyone would want to represent or pay for the right to view what you've made.

Now this situation doesn't mean you need to be exploitative. Any quick glance at a video store shelf will reveal the public's purported interest in guns, girls, and recognizable movie stars. Most video packaging features some combination of these elements. And, quite possibly, *those* are the best identifiable elements to use in promoting your film or video (at least the guns and girls, since you won't have money for the stars). However, there are lots of projects out there that don't have such easily identifiable marketing elements, and with these, you will need to work a little harder to find a commercial angle. Feature-length films are the most susceptible to this problem. Instructional videos and documentaries have built-in promotional angles. There's not much mystery in finding the selling points of a production entitled *How to Lose 5 Pounds in 5 Days*. The same goes for a public or cable television special dealing with the erosion of the traditional American nuclear family. Quantifiable audience segments for these kinds of projects are known before the script is finished. What do you do, however, with a 90-minute fictional film called *Open House*, that's a sorta-funny, sorta-serious independent feature about four housemates and their daily struggles? It doesn't fit into any easily promoted genre (horror, western, suspense, comedy), and it doesn't feature any well-known actors. How do you find an audience if all you promote is the artistic merit of the project? How do you interest a buyer if she can't understand the selling angle? If you're unable to promote the potential commercial worth of your product, it's going to be tough going on all fronts. Whether you are trying to attract consumers, distributors, or the media, you need to find an element that allows for an immediate connection with them in some familiar manner. And once that is accomplished, you will basically promote the hell out of that one element.

The arsenal of tools available at a film promoters' disposal for this task is vast. From direct mail and personal sales to custom Web sites and trade magazine advertising, there are multiple ways to bombard potential viewers or buyers with your "angles." But how, and when, do you start this whole process?

START AEAP[2]

Before we get into the "how," it's important to learn the answer to the question of "when." Now. That's your answer. If there is one bit of

---

[2]AEAP stands for As Early As Possible. Many years ago, one of my filmmaking teachers told me that when it came to independent filmmaking, I should start telling people about my movie as early as possible, because it would help with financing if there were greater awareness of the project. His theory was that if you told enough people you were making a film, sooner or later the project would take on a life of its own. Though we were discussing ways in which to locate project funding, I truly believe the promotional aspect of the film and videomaking process can benefit from the same logic.

knowledge I want you to glean from this book, it is to begin promoting your independent film or video project the day you decide to make it. I'm going to repeat that thought so you can read it again:

Begin promoting your independent film or video project the day you decide to make it.

You are probably wondering how can you promote something you don't have? Very easily stated, you will market your *idea* until a tangible object can take its place. This process happens with every product, every service, everything bought and sold in the world today. How do you market an idea? Let's consider an example:

Titanic was, in 1997, the biggest, most expensive, and most hyped movie ever made. Do you think there would have been such public interest and anticipation for the project if producers hadn't begun marketing and promoting the *idea* of the film months before even one foot of film was shot? Director James Cameron had the *idea* to create the biggest and most expensive movie ever, and his promotional team took this *idea* and let the public know about it. The awareness rate for *Titanic* during its development stage (read as *idea* stage) was greater than for most films playing in theatres at the same time. More American people knew about the *idea* than they did about the finished product!

While *Titanic* may be thought of as an out-of-reach example, the basic principles used to create the success of this multi-million-dollar Hollywood blockbuster can be utilized on even the lowest budgeted film or video project. Creating public awareness, public interest, and public anticipation for your *idea* are all attainable goals that any promotional plan should address. These are the basic blocks that will build your *idea* into a saleable product.

Sure, production value, recognizable talent, breath-taking cinematography, and a polished script help your cause. But the reality is a film properly promoted and marketed can succeed even without these elements.

Case in point: my very first feature film project, *Killer Nerd* (made with Wayne Harold, now president and owner of Lurid Entertainment —

www.lurid.com — a film and web production company. Harold was also co-creator of *Bride of Killer Nerd*, and his first solo film, *Townies*, has been released as a special edition DVD from independent specialty distributor Tempe Entertainment). The movie was made on a budget of under $20,000 with amateur actors, real-life locations (such as friends' homes), and homemade special effects. While it's better than my Uncle Joe's home movies, it's not going to win an Academy Award for technical achievement anytime soon. However, through an exhaustive and creative promotional campaign (one that did not cost thousands of dollars), the film was hyped for weeks before its release on MTV, CNN, *The New York Daily News*, and countless other regional newspapers, magazines, and radio and television stations. The movie ended up on video store shelves nationwide; was sold in foreign markets like Malaysia, Mexico, and Sweden; and even made it to the retail shelf of several large mass merchandisers. We knew who to target for our promotional mailings, what type of publicity stunts to pull, and when and where to release the film to guarantee the largest amount of public awareness (and sales). It all depended on finding that one promotable element — in this case, the star of the film was truly a nerd in real life (more on that later) — and telling the same story over and over to anyone who would listen.

But before you start promoting anything, before you pick that element of your film or video that's sure to lead you to Hollywood, before you start dreaming of limos and lap pools, you need to ask yourself several questions about your project. You need to get intimate with what it is you are really working on, and be ready to do some serious thinking so you can clearly determine answers to the questions of what do you have and who wants it.

Chapter 1 Summary Points

- Film promotion is just as important today as it was in the dawn of the entertainment industry.

- Producers working within the studio system get paid for their efforts whether their project successfully finds an audience or not.

- Most independent producers will not work within the studio system — at least initially.

- The level of commitment to a promotional campaign cannot be used to forecast that project's ultimate success.

- As with all products, a film's marketable elements must be identified *before* the promotional campaign is created.

- It's important to begin promoting an independent film or video as early as possible.

WHAT DO YOU HAVE?

Sure sounds like a simple and stupid question, doesn't it? Well, if your response to this simple question goes something like, "My movie starts off with this spaceship landing in the middle of a high school football field during the championship game and a huge alien gets out and starts abducting cheerleaders, but it's funny because the alien is wearing a dress instead of a spacesuit and one of the football players is attracted to him...," then you better sit down and give it more thought. The above-described film begins as science fiction, turns to horror, but finally sounds like a black comedy! This sort of description will only succeed in turning away everyone who enjoys any one of the individual genres your film is trying be. One fact to keep in mind:

Your movie or video project cannot be all things to all people.

You should be able to define your film in one concise sentence, identifying its genre in the process while at the same time conveying the basic storyline to the potential viewer. This method of summarizing your entire film into a short sketch is known as the **concept capsule**. Let's consider a basic example:

King Kong is the exciting story of a giant prehistoric ape, stolen from a tropical refuge, struggling to survive in modern-day New York City.

The above sentence identifies the major storyline (stolen from home, forced to live in a foreign land — the classic "fish-out-of-water" scenario), describes the settings (tropical refuge, New York City), and even includes an emotional plea (struggling to survive).

Creating a concept capsule allows you to focus on what your film or video is really about. This exercise forces you to unearth the heart of your story, exposing topics and themes that will become the basis of your entire promotional campaign.

Concept capsules do not come in a flash. Creation of a fitting capsule requires you to do some serious thinking about what it is you are really working on. You will undoubtedly come up against mental blocks, causing you to believe it's impossible to boil your opus down to a single sentence. It can, however, be done by working through a couple of exercises.

Before tackling these project-defining exercises, go to a quick-copy center and get some basic business cards printed. Several hundred will cost less than $15. On the card, put your name, the project title, and a phone number, e-mail address, or street address where people can reach you. Don't ask — just do it. You'll need these soon and you want to have them ready before you get too far into this activity.

CONCEPT EXERCISE 1 – CAPSULE BREAKERS

For starters, jot down answers to the following "capsule breakers." These questions will get you thinking about your project in new ways:

1. What is the basic structure of your project? Consider length when deciding if it's a feature-length film (70+ minutes), documentary, instructional or informational tape, short, or something else.

2. What is your project's prevailing genre? Would it be categorized as science fiction, horror, western, drama, action, children's, erotic, reality-based, comedy, cult, anime, instructional, educational?

3. What are the project's emotional pleas? Will the viewer simply be entertained, or is the purpose or effect of the project to enlighten, educate, inform, sadden, madden, scare, or elicit pity? Can the viewer relate to the information or material in some personal manner? Is the project experiential?

4. Does the film or video make a call to action? Is the viewer prompted to buy something, change his way of thinking, alter his lifestyle or daily habits?

5. Who are the principal characters within the parameters of what is being presented? Are they fictional or actual beings? Are they historical, present-day, or future beings? Are they human, animal, or other-worldly creations?

6. What is the major action? Is the viewer presented with a hero's journey, a tale of forbidden love, search and conquest, reversal of fortune?

7. Does the project deal with a current event, fad, or popular culture phenomenon? Is the subject popular with specific audience segments? Will the time frame of the project's appeal be limited?

8. Does the film or video depend upon realistic special effects or stunts? Are you showing off new technology by painstakingly recreating a time or a place through the magic of visual effects?

9. What are the physical and time settings? Does the project showcase never before witnessed landscapes? Are you exploring fabled lands? Is the setting past, present, or future?

10. Who is the hero, the protagonist? Who drives the action of the story? Is there a love interest? Is it believable? Are they typical for the story? Is their gender important? Will the audience relate to these characters? Will viewers like these folks?

11. Why does the audience care? Who would be attracted to your project? Are you a member of the target audience?

Certainly this list is not exhaustive, but it gets you going in the right direction.

Don't expect to answer all of these questions in one sitting. Take some time to think thoroughly about your responses. If other people are responsible for the project, ask for their input as well. Once you feel satisfied with your answers, it's time to move on to a second exercise that will help you more accurately define "what you have."

CONCEPT EXERCISE 2 – PERCEPTION VS. REALITY

This exercise is really a two-step process. First, for ten minutes write down as many words as possible that you feel describe your film or video. Don't think too much during this process, it's important just to get your ideas on paper. Keep track of all ideas, no matter how weird or inappropriate they may seem.

After you have finished making this list, review your project. If you are at the script stage, reread the script. Those with a finished tape or film, it's time for another viewing. And for the folks just in the planning process, look over any notes, drawings, or other items you have associated with your project.

Often, there is a disparity between what you *think* your project is about and what it truly represents, and comparing your word list to reality is a neat way to extract these perceptions. After reviewing the project for a second time, add any new or more complete descriptors to your list that might have been missed initially. Do not remove any words or phrases that seem purely imagined, as often these become key to your final promotional campaign (more on this in a bit).

Now take a deep breath, clear your mind, and choose three or four words (no more) from your list that best describe your project in a visual or emotive manner. These can be words that identify feelings evoked by the film or actual descriptions of the action taking place, just be sure that the words selected are those offering the most "heat."

Let's look at an example. When beginning the official promotional plan for the film *Pig*, we had a script, some concept art, a logo, and about 200 location photos. We did not, by any means, have a finished project. However, my partner (T. Michael Conway, president of Paragon Pictures and writer/director of the horror feature, *June Nine*) and I had a very clear (previsualized) idea of what the finished film would look, feel, and sound like. So, after "watching" the film in our heads, and reviewing the script, location photos, and other production planning materials, we came up with the following list of words:

> Police
> Violent

Bored
In-your-face
Kids
Religious
Local
Handheld
Tough
Bad ass
Mean
Hidden
Reality
Small-town
Racism
Sunny
Warm
COPS
Back roads
Lonely
Punks
Melancholy
Independent
Rural
Slice-of-life
Drifting
Tired

As you can see, this exercise really stretches the brain. You'll find words popping into your head (like warm) that you would never consciously associate with your project. It's a good way to get more in tune with the "feel" of your project, and these "feelings" are what promotion is all about. Before anyone actually sees your film or video, they'll be exposed to your promotional campaign. This campaign can only be successful if it "feels" right to the audience.

How many times have you gone to a movie and been disappointed because the product did not live up to the feelings (often referred to as **hype**) created through the promotional campaign? Sure, sometimes it is a simple case of false advertising — that action-film-of-the-year only features one car crash. Usually, however, the mismatch between audience expectations and reality is due to the promoters not thinking through this step. There's something to be said for using a good **hook** to reel in audiences. You'll sell a lot of product and make quick money. But what happens when you try to sell Project #2? Chances are good you won't, because all of those "suckered" buyers will be on to your empty promotional promises. Remember, your film or video can't be all things to all people. Keep that in mind when choosing those words that will be used to craft your concept capsule. Use only those words that truly represent the project to its intended audience.

Back to the example; here are three words we chose and why:

1. *COPS* – Everyone is familiar with the television show and the film would be shot and edited in much the same manner.

2. Violent – Though the movie contains only four scenes that would be characterized as excessively violent, there is a *feeling* throughout the entire film that creates a sense of tension and a sense of dread. The main character always seems on the edge of violence, which literally paints the whole film as violent.

3. Rural/Small-town – *COPS* is always associated with large metropolitan areas, giving viewers a taste for what a big-city police officer's life is like. The whole premise of *Pig* was to explore the life of a small-town cop, a character that most people in America can identify, but know nothing about.

Now, do the same. Look for those words on your list that provide the most insight, the most information about your project. Think about each term and how it's relative to both the project and the feelings you want the potential audience to experience.

With our three words chosen, the following concept capsule was developed:

Pig is a *COPS*-like look at the violent life of a small-town police officer.

The sentence fulfills the requirements set forth earlier by describing the setting (small town), action (police life), characters (police officer and people he encounters), and emotional pleas (the promise of a voyeuristic look into the life of a small-town police officer) of the intended project.

WRITING THE CAPSULE

Now take your words and weave them into a sentence. Rewrite that sentence until it "tells the story" in as few words as possible. It may take you a day or two until you are satisfied. After you've settled on what you believe is a good concept capsule, write it down on a 3" x 5" card and circulate it among people who are completely objective to the project (you can try family and friends, but they probably won't tell you the truth if it's negative).

Online film discussion boards are really a good place to work through this step. You also will want to try eliciting responses from individuals without a vested interest in film or videomaking. Again, people impartial to your goals — complete strangers, if available, are the best — provide superior feedback. Coffee houses, classrooms, and other relaxed gathering places where you are in close proximity with strangers are ideal settings for testing your ideas.

If you're planning a project for an extremely specified target audience (i.e., an instructional tape on easy ceramic tile installation), solicit this group (in this case, home remodelers) for input. It's vital that the core viewer understands your promotional campaign, especially if they represent a small or limited number.

Just as with most of the exercises and techniques set forth in this book, you're going to have to insert yourself into situations requiring interaction with people you've never met in person, never spoken to on the phone, never communicated with via electronic means... basically strangers. Starting conversations with people with whom you share a common experience (taking a night class together, sitting in the same neighborhood café, etc.), gives you good training when the time comes to approaching total strangers. You need to overcome any innate fear of talking to unfamiliar people, as networking and personal connections are the basis for success with film and video marketing.

Ask people what they think the described project is about and if it is a film or video they would be interested in watching or buying. If the comments concerning your concept capsule match what you feel is the heart of your project, great! If they don't, find out what aspects people are misunderstanding and begin again.

Be sure to keep copious notes on all responses and the respondents themselves. You may find that specific segments of the general population react differently to your idea.

When gauging reactions for the *Pig* concept, we found that many people felt the planned movie was going to be scene after scene of police officers chasing, beating, and arresting lawbreakers. This was not the case. Another conclusion from the survey was that many females, especially those over 35, did not seem to have much interest in viewing such a movie.

The first problem was easily solved. We went back to my list of words and chose one to integrate into the concept capsule:

Melancholy – a big portion of a small-town police officer's daily schedule is filled with eventless, hour-upon-hour driving and sitting – waiting for action. I wanted the potential viewer to know this aspect would be investigated in the movie.

Thus, the capsule now reads:

> *Pig* is a *COPS*-like look at the violent and sometimes melancholy life of a small-town police officer.

As for the second issue, I didn't think many women in that age group would be attracted to a film called *Pig* that dealt with police life. The film wasn't targeted at that demographic. *Pig* was aimed at a predominantly male audience, and especially a young male audience. I knew that *Pig* couldn't be all things to all people, and it would only have negatively affected the campaign if we changed the capsule to make the project sound more attractive to females over 35.

Remember those cards I told you to make at the beginning of the chapter? Well, now is the time to hand those out. After thanking the respondents for their help, give them the card. It's a very, very cheap way to begin promoting your film. I guarantee the person you question and leave with a card will tell at least one other person about the experience. Some will think you are weird, but most folks will appreciate your sincerity and motivation to accomplish such a seemingly monumental task as making a movie or a video. People don't run into a film promoter every day, and it's that uniqueness that will get your project some initial attention.

Filmmaker Kat Candler used business cards beginning at the earliest stages of her project Cicadas.

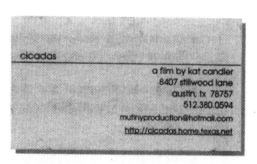

cicadas

a film by kat candler
8407 stillwood lane
austin, tx 78757
512.380.0594
mutinyproduction@hotmail.com
http://cicadas.home.texas.net

I know this seems like an awful lot of work just to write a sentence that describes your project. It is humbling and time-consuming, but it is a process that must be carried out correctly. Though you'll likely become frustrated when people aren't "getting" the idea of your film or video from your concept capsule, this exercise is actually quite beneficial. It allows you to step away from your film and look at it through an outsider's eyes. Often, we get so close to a project we just assume everyone will know what it is we are trying to do. Not so. The whole idea behind film and video promotion is to create a campaign that will attract these outsiders to your project. Constructing the concept capsule is the first step (and possibly the hardest) in building a successful promotional plan. Once you have your capsule, it's time to discover who wants to swallow it (sorry, pun intended).

Chapter 2 Summary Points

- Don't make the mistake of believing your project can be all things to all people.

- A concept capsule concisely defines your film or video's genre and storyline to the reader.

- To write a comprehensive concept capsule, you must consider your project's genre, emotional pleas, characters, action, and settings.

- Brainstorming allows you to free-associate feelings about your film or video, leading to a clearly descriptive concept capsule.

- Continual refinement and close investigation of every word in your capsule is necessary to ensure it accurately reflects your project.

- Try a concept capsule out on various audiences to be sure it is "saying" what you want it to say.

WHO WANTS YOUR PRODUCT?

The title of this chapter might seem like another dumb question, right? The way you see it, the question poised should be: Who doesn't want your product? But, that's not always the case. Let me prove my point.

Before you read any further, write down the names of ten people who you believe would spend $30 to buy your finished film or video.

Now, considering your list, answer the following:

- Do any of the names on your list represent people who can persuade others to buy your film?
- Did you name someone with the capacity or desire to buy more than five copies of your finished tape?
- Could anyone on your list be considered an influence over a large group of people?

Well, if you answered "no" to those questions, then your list probably resembles one of my first stabs at this exercise, and includes the names of close friends, family members, and others personally interested in your film and video pursuits. To be totally honest, I actually had trouble coming up with ten people off the top of my head. I mean, I *knew* there was a market for our film, I was just having trouble specifically identifying the people in it. Not a great way to begin promoting a project that represented both a lot of personal time and money, not to mention something on which I was hopefully basing my future.

YOU CAN'T SELL TO STRANGERS

There's an old salesman's saying that goes, "You gotta know 'em to show 'em." What this means is that you must be able to identify an audience before you can present them with your product. Basically, if you don't know who is interested in your film or video, and why they're interested, you're going to have a tough time when it comes to selling the thing.

Telemarketing is a great example of this theory. Those annoying calls-in-the-middle-of-dinner often miss their mark. And it's not just because they occur at the worst possible moment. Usually the solicitors on the other end of the line don't know the audience. Some obviously conduct rudimentary research, such as replacement window salespeople calling anyone who has recently secured a home improvement loan. But more often that not, you'll get a call for a product or a service that you have absolutely no interest in purchasing. You don't even want to hear the pitch.

On the other hand, sometimes the telemarketer has done her homework. For instance, I subscribe to several fitness magazines, and in the past couple of years have purchased a fair amount of home exercise equipment, pieces that all were warranty registered through the manufacturer (which gets the buyer's name added to solicitation lists). I recently got a call offering me a subscription to a new health magazine that came with a free yoga video. The magazine was specifically oriented to the home health enthusiast, not those folks who frequent the gyms and juice bars. Of course I listened to the salesperson's plea and of course I bought the magazine.

Now, this example also provides a good analogy to your position in the industry as an independent producer/promoter. The telemarketers calling anyone and everyone (in this case, the replacement window manufacturers) are like studios. With what they believe to be a mass-appeal product (everyone needs windows) and the budget and resources to back such an effort, these companies have the potential to succeed just through the sheer number of sales attempted by their staff. This type of telephone promotion campaign can be likened to a movie advertised nationally on primetime television, in major magazines, and across the Web before opening on 2,500 screens.

The small, start-up health publication to which I subscribed is more like the independent producer. To succeed, they must carefully select a target audience and promote specifically to that group. Without a large corporation (read studio) covering their back, each dollar spent and every hour dedicated to the sales effort must return a profit. The health magazine publishing company, like a first-time film promoter, doesn't have the luxury of realizing economies of scale that come with national sales staffs, and must work smarter and harder to obtain sales.

YOUR THREE AUDIENCES

When independently promoting a film or video, properly identifying your audience is a little more difficult than singling out the typical replacement window or magazine buyer — because you really have three separate audiences, not just one. They are:

- Those who will watch the product (consumers via purchase, rental, or broadcast situations)
- Those who will buy/sell the product (distributors, acquisitions agents, subdistributors, video buyers)
- Those who will promote the product (publicists, media, festival programmers)

Let's attack each of these separately.

FINDING THOSE WHO WILL WATCH — ANSWERING THE "BIG 11"

With your concept capsule in hand from the last chapter, think about your film. What elements does it contain that you know will appeal to specific segments of the population? Is it a love story? Then certainly women will top your list of potential viewers. But what if the love is expressed between two 14-year old kids? Has your viewership changed to teens?

When developing a promotional plan, the identification of your film or video's potential viewing audience is, without a doubt, equally as important as creating the concept capsule. With the latter, you described what product you will market to the public. Now it's time to describe that public.

In order to obtain the clearest picture of who will most likely be attracted to your project, it is necessary to create an **audience profile**. This exercise allows you to distinguish common traits and characteristics found in and exhibited by your potential viewers. Creation of the audience profile begins by asking yourself the following 11 questions concerning your film or video:

1. What is the primary gender of your audience? Males will go for action films, females like dramas with a love interest, and comedy appeals to both sexes. While there are no hard and fast rules here — many women loved *Gone in 60 Seconds* and *The Fast and the Furious* — you can usually make a pretty good judgment by examining the basic content of the story. This process may at first seem like stereotypically drawing conclusions (with this and many of the following inquires), but research of film viewership as determined by sex and genre backs up such a conclusion.

2. How old is your audience? Will the film or video appeal to children, teens, young adults, middle-aged folks, or seniors. Some films will cross all age lines (*E.T. the Extra-Terrestrial*) though many low-budget, independent feature films are targeted at the 18-34 crowd. Consider violence, language, subject matter, talent, and production values of the project as indicators of age-level interest.

3. What is the educational level of your audience? High school, college, graduate-level, or higher? Most sophisticated period dramas and "talky" movies appeal to a higher-educated crowd than "shoot-'em-up" films and home repair education videos.

4. What is the income level of your audience? Lower income individuals (not counting college students and "slackers") tend to avoid talky dramas, while individuals in the upper tax brackets will usually forego instructional tapes (believing it is easier to hire someone to do the job).

5. What is the religion of your audience? Though not as prominent a consideration as it was in the past, devout Catholics and Baptists may still avoid anything overtly violent and/or containing excessive sexuality and lewd language.

6. What is the race of your audience? Again, not a prevalent issue with most projects, but I tend to believe urban-themed films play better to an urban crowd.

7. What is the occupation of your audience? White-collar workers may enjoy courtroom dramas, while the blue-collar crowd is ready for *WWF Smackdown II*. Medical and educational professionals might be interested in a cable program on advancing electronic technology for the home, and automotive technicians (the politically correct term for mechanics) are probably better suited for your video entitled *Profiting from Swap Meets*. Contrary to what some might believe, however, occupation and income level many times have little correlation.

8. Where does your audience live? Rural residents may not be too interested in a video detailing subway self-protection tips, whereas urban residents probably won't be the first in line to see a documentary on building a home-based livestock stable.

9. What are the habits and interests of your viewers? Heavy readers go for Woody Allen-type movies. Sports enthusiasts like action flicks. Television viewers like a wide range of films. And do-it-yourselfers are interested in, believe it or not, do-it-yourself videos.

Dragon and the Hawk was made by and for those with an enthusiastic interest in the martial arts.

10. What are the purchase patterns of your audience? Impulse buyers may watch and buy any movie on a whim. Those more careful with their funds will closely consider the content of a film or video before making a decision. This is a very important consideration if you are promoting a "purchase-only" project such as an instructional tape.

11. What are the major motivations of your audience? Are they politically involved? If so, with what party or cause? Are they feminists, chauvinists, or sexists? Do they believe in the work ethic or are they daydreamers?

You are not trying to stereotypically profile your audience even though the above questions probably make it sound that way. You are simply trying to get the best "read" on those folks who are most likely to view your product, and by demographically dissecting the entire pool of potential viewers (basically everyone), you are able to paint a very specific picture of your intended viewing audience.

CONSTRUCTING THE AUDIENCE PROFILE

After carefully considering the answers to each of the inquiries, compile your responses into a cohesive description, one that should look something like:

Pig: Audience Profile
Males, 15-35, low-to-middle income, high school and college educated, frequent television viewers, sexist, possibly racist, attracted to sensationalistic current events

Now ask yourself the big question: Looking at the audience profile, will potential viewership be large enough to warrant production of the project? Well, if your film or video is already completed, this is pretty much a moot point. No matter how small your intended audience may seem, you can always come up with a promotional plan that will help you locate viewers you may not believe exist (and no, it doesn't include lying). But if you are still in the preproduction phase of the project, you possibly just saved yourself a lot of time and money. Recognizing an inadequately sized audience affords you the option of rethinking the whole project, or maybe just changing some aspects to attract a larger crowd.

After creating the *Pig* profile, we realized a sizable audience existed for the kind of film we were producing, partly because my partner and I were members of that group. Since guys like myself rent a hell of a lot of flicks and have pushed reality-television programming, not to mention reality-based movies, to the tops of their respective charts, we were assured a sizable audience base existed — one large enough to warrant production.

Realization of an assured audience actually *prompts* the production of many films, so this kind of profiling exercise often takes place before one word of a script is even written. Prolific low-budget auteur Roger Corman built a filmmaking enterprise through the utilization of such techniques.

Starting with a movie title, or more commonly, a film idea very similar to something already popular in theatres, Corman could guarantee some level of success if his budget and audience size estimates stayed within reasonable parameters. For example, after Steven Spielberg gave the world *Jurassic Park*, Corman quickly shot and released *Carnosaur*. The Corman project was one of his most expensive films ever, but he knew a huge, worldwide audience was left wanting for more after the record-setting Spielberg film came and went. With almost identical audience profiles, Corman knew he would have a hit on his hands as well.

In contrast, the Ice Cube and Jennifer Lopez-starring *Anaconda*, though not a flop, certainly wasn't a huge money-making franchise. Corman again sized up audience potential through profiling and offered *Python* as his addition to the reptile movie craze (which is still being played out: Witness Columbia TriStar's latest direct-to-video feature, *Boa*). Not nearly as expensive as his dinosaur opus, Corman's *Python* is very profitable because the budget was partially dictated by potential viewership.

This is not to say that extreme niche products cannot succeed, such as many instructional tapes, documentaries, and even some feature-length films. The size of the audience in these situations must be considered in light of the project's budget. A square dancing competition highlights video made for $2,000 is not that risky. Though not a huge segment of the population, enthusiastic pockets of these dancers exist in most every state and can be reached through a number of different specialized media. However, if such a narrowly targeted video was produced for $20,000, you may have a problem recouping your costs.

PLAYING BOTH SIDES

While it's true your movie cannot be all things to all people, you may discover while building your audience profile the emergence of two distinct groups of viewers. *Blue Velvet* was initially marketed to the college-age group based on director David Lynch's reputation for dealing with the bizarre in a highly imaginative fashion. It was also promoted to older, intellectual adults, film enthusiasts, and critics who raved over the film's superb story, acting, and direction. This is not uncommon with many so-called independent Hollywood films. *Pulp Fiction* is another good example. One campaign featured the violence and snappy dialogue many expect from the film's director, Quentin Tarantino. Another promotion centered on the comeback of John Travolta (including his famous leg work) and was obviously created to attract the female viewers who weren't into the blood and guns so prominent in the Academy Award-winning movie.

If you start to see this pattern developing in your profile, you may want to consider two different promotional campaigns aimed at the two different groups of viewers. This procedure is expensive and difficult to coordinate. The effort to create advertising, logos, promotional art, press releases, photographs, media kits, interviews, publicity stunts, Web sites, sales approaches, target markets, and everything else that goes into promoting a film or video may likely have to be duplicated to ensure that you are clearly communicating your main message to each of the defined audiences.

I would only suggest this approach if the added audience (often secondary in size to the initial one) is large enough to warrant the extra time and expense involved in developing a new set of campaign materials. You surely don't need to be reminded that you are working with a limited budget where every penny counts. Usually it's best to spend these funds on a single identifiable group. But your project may be an exception.

Another caution when working with dual campaigns: setting expectations that might not be met. Let's say you have made that science fiction-horror-comedy described earlier in the text (you know, with the cross-dressing, co-ed abducting alien). As a "beer-and-pizza" flick, your first audience is going to be college kids looking for a fun movie to pass the evening. You also may believe the movie has a nostalgic sense to it — harking back to the drive-in days of the 50s and 60s. Thus you create another promotion to lure baby boomers and others who grew up in that era. This secondary audience will not respond well when the movie deviates from its promise of "good ole-fashioned fun" by showing graphic gore, sex, and language piled on top of a modern rock and hip-hop soundtrack. Bad word-of-mouth from any targeted audience could hurt promotional efforts aimed at other groups.

Information contained in your audience profile should get you thinking about promotional techniques ("hooks," artwork, media, etc.) best suited for your project's potential viewers. For example, it's pretty obvious that the viewers of the *Cheerleader from Mars* movie won't be basing their rental decision on a bunch of critic's quotes. A video box featuring a scantily clad cheerleader standing next to the alien would most likely work best.

While not all product will be so easy to represent through art, profiling your audience by answering the 11 queries listed earlier gets your creative juices flowing. As promotional ideas come to you, keep a notebook specifically dedicated to that purpose. Nothing is dumb at this point. Write everything down, no matter how outrageous or silly it seems. Before we get to putting all those thoughts into a campaign, however, you need to consider your "other" audiences.

BUYERS AND SELLERS

As a promoter of filmed or videotaped entertainment, you are going to have to do business with a lot of people of varying abilities, professionalism, experience, and — for lack of a better word — clout. **Makers** and **breakers** is probably a more accurate description. These are the folks who can take your project and make you rich and famous, or forget that your tape is sitting on their desk and refuse to return your calls. Regardless of the outcome, you need to identify this audience of buyers and sellers, as often they are more vital to your success than the ultimate viewer.

I know, I know. I just told you that identifying the intended viewing audience for your film or video was a very important step to your promotional campaign's success. And that's still true because it allows you to better understand your project from the audience's point-of-view. That knowledge will come into play in every situation you encounter from here on out. But that alone will not get your deck-building instructional tape on the shelves of Home Depot, your murder-mystery movie programmed on HBO, or your animated short touring the festivals. Knowing the players and the channels that lead to those outcomes is the only way to realize those outcomes.

In addition to buying and selling your product, many of the following folks will also work to promote your film or video. Not in a traditional, public-based manner such as through a critical review or a radio interview, but in the business-to-business sense — whether that means a distributor wines and dines a big video chain owner while explaining the merits of your latest movie, or a producer's rep works to convince a film reviewer to include an analysis of your film in his next column. In the entertainment industry, there are no hard-and-fast rules as to what a person's specific duties include. It gets down to finding success in the marketplace for whatever product he is representing and using whatever means necessary to get that done.

Distributors
Can't live without 'em. In one way or another, promoters of all but the smallest, most niche-targeted projects will deal with some kind of a distributor, whose sole reason for existence is to get your film or video in front of an audience. Whether it is a rental, purchase, broadcast, or other viewing situation, a distributor is involved. Distributors buy and sell and promote any and all kinds of movies and tapes, often working with other distributors to get the product to the final viewer. These big cats so command the entertainment market, they are described in intricate detail in Chapter 7.

Producer Representatives
Depending on your project's status (preproduction, production, post), you may have already spoken with a producer's rep. Often scouting out films in their earliest stages, producer reps will work on making inroads for your project with distributors, the media, agents, publicists, and the like. Their job is somewhat nebulous and hard to define in concrete terms. Basically, they are facilitators and communicators, making connections for you and your project to ensure a wide variety of objectives, including project completion, maximum media coverage, wide distribution, sales, and the ability for you to create another film or video.

Director Michael Mongillo suggests independent filmmakers seriously investigate the use of a producer's rep as he did with his flick, The Wind. "A good producer's rep knows more about the buying and selling of a movie than most independent filmmakers know about making one. A rep can get your film into the hands of the decision-makers at distribution companies, not their assistants." Mongillo used this flyer to attract initial distributor and rep interest in his film.

Love comes in many forms.

Unveiling:

Angelika Film Center
Theater 5
Thursday, September 23, 1999
4:30 PM

www.thewindmovie.com

Distributor Representatives

Functioning much like a producer rep, distributor reps largely tend to making distribution deals, though they handle other functions such as media relations as well. These guys only work with projects they strongly believe have a chance of "making it," for the simple reason they won't make any money if it doesn't. Distributor reps will also involve themselves as early as possible with a project, hoping to guide it in a manner that best suits possible buyers. Distributor reps lend a lot of clout to a fledgling film or video, especially so during the important selection process of prestigious festivals like Sundance, Slamdance, and Telluride. You can rent-a-rep (the same goes for producer reps), essentially paying for their services and connections. But don't expect fame and fortune overnight. Hired representatives are often working with many producers, giving each cursory treatment at best. The ideal situation is to find someone who is as passionate about your project as you are.

Agents

Too many bad things have been written and perpetuated about agents. Hollywood is a strange place, and agents seem to take the brunt of all that strangeness. Plain and simple, agents make connections. Just like Tom Cruise pushed to get Cuba Gooding Jr. more money and more fame in *Jerry Maguire*, entertainment industry agents do the same for their film and videomaking clients. Agents handle many of the same functions as producer and distributor reps, and even publicists, but usually with a higher level of connections. They have to, and do, know everyone. Even at the lowest levels, a good agent will be familiar with everyone up the food chain. From video buyers and cable television acquisitions officers to packaging artists, musicians, and all pertinent media, agents are connected. Their usual day may include hundreds of phone calls and 10 meetings, as their livelihood depends on making and keeping personal relationships. Agents work on commission or work for free if they feel passionately about a project (or at least are passionate about how much the film or video might profit). Agents love the **sizzle**, the promotional aspects of a project, and will react accordingly to a thoughtfully presented plan. Just as with distributors, there are many levels of agents, each vying for the hottest, most attractive new films and videos.

Home Video Buyers

As an independent producer, the home video (and DVD) market will likely become your holy grail. That's because most low-budget productions will forego any theatrical distribution, instead heading for a direct-to-video release pattern. Depending on your project, this means you'll want to befriend either rental or retail buyers, or both. Both are solicited by national video distributors, so those readers securing distribution contracts won't get too involved with this group. Regardless of your situation, it is always helpful to know who is buying your product, especially since this segment accounts for most independent filmmakers' largest sales potential.

There are two types of buyers: rental and retail. They can represent a national or regional chain, a single independent store, or anything in between. The buyers themselves are some of the most difficult-to-contact folks in the business. That's where the similarities between the two end, however. Rental and retail buyers have very distinct needs, driven by their size, location, merchandise, and markets.

Buyers for a video rental chain face the dilemma of **breadth vs. depth**. Should they stock 100 copies of the latest blockbuster or stock 20 units each of five recent releases? As an independent producer/promoter, you'll obviously have better luck with those buyers looking to add variety to their store shelves. With retailers, breadth is the rule, except for the biggest Hollywood hits, which will be stocked deep.

Size of the chain is another issue. Believe it or not, the smaller the rental chain, the more likely they are to buy your tape. The reasoning here is that corporate monsters such as Hollywood Video, Blockbuster, and Movie Gallery handle so many tapes, that for them to do business with a single title supplier (this being you), would require an enormous amount of manpower. Just as you don't want to go to a light bulb store, a cigarette shop, and a fresh milk outlet when you can get all of these simple household items in one place, neither do the big video chains.

Retailers selling video provide a much tougher nut to crack. While some retail shops provide an environment conducive to selling entertainment product (Borders, Best Buy), those who hawk everything under the sun will be more selective in what they stock. Large specialty or niche retailers like Home Depot and Bed, Bath & Beyond will stock programming targeted to their core customers, with small local book, drug, and record shops likely to carry anything with a local connection.

Ratings restrictions, content issues, video box art, payment terms, and delivery options are all issues requiring much less bureaucracy within smaller chains and individual retail stores, too.

Pricing is the sharpest divider between the rental and the retail buyer. Video stores, be it your local mom-and-pop down the street or the neon-splashed superstore consuming half a city block, are going to pay more for your product than any retailer. The rental buyer knows that a tape will continue to generate revenue for at least six months or more. He understands the value of the product.

Retailers, on the other hand, view your product like anything else on their shelf — as inventory. The tape cannot be priced too high or the consumer will just rent the film or video instead. This **sell-through** pricing level — where the low cost entices consumers to own a movie — usually hovers around $15, though many old movies and video programs can be found for as low as $4.99. A good example: my experiences with *Killer Nerd*. We sold the movie to rental buyers for anywhere from $20 to $30. When it came time to solicit retail stores, however, the most we could command for a copy was $7. Understanding pricing issues is one of the most important aspects to successfully working with rental and retail buyers.

Cable, Pay-Per-View, and Broadcast Television Buyers

You've been living in a hole if you haven't been contacted to buy some sort of expanded cable, digital, or satellite-based television system. With more than several hundred selections available on most of these systems, even the most arcane interests and hobbies have a channel, or at least a couple of programs, dedicated to their pursuit. And that's just counting the U.S. market. Once you add up all of the networks worldwide, it's easy to understand why this has become such a lucrative market for independent producers. It is a simple matter of supply and demand. Too many channels plus too few shows means good news for those film and video promoters with suitable product.

Television audiences are fickle folks, a fact well understood by programming executives. Thus, the acquisitions agents responsible for filling cable, pay-per-view (PPV), and broadcast schedules are always on the lookout for new, dependable suppliers. With cable and PPV, the parameters of what is acceptable, and wanted, is fairly diverse. Cuisine, home repair, nature, and history are easy targets. But glance through a *TV Guide* and it's not uncommon to find listings for automotive shows, ballroom dancing galas, homeopathic health instruction, and educational programming of all kinds. When you consider leased and local cable access channels, your options are even greater.

Broadcasters like Fox, NBC, and WB hold the reins a lot tighter; there's not much room for independent fare on these networks, unless you luck into local network time (usually available late, late night or early Sunday mornings). Local broadcasters on the UHF band (higher than channel 12) and low-power television (LPTV) are a different story. Buyers look to a wide variety of sources to fill time slots, and independently produced video programming regularly appears on these channels.

Technical quality will be the biggest drawback for most low-budget producers, but buyers in this category usually have a well-developed and easily accessible set of standards that describe programming needs, including running times and content constraints, in strict detail. One final note: Most broadcast and cable buyers will consider series programming before one-shot shows.

Internet Buyers

What started as a way for people to exchange digitally encoded information has ended up changing the way we live, and the Internet has had no less of an impact on the world of independent films and videos. Opportunities of every kind exist through the Web, including online sales, streaming, critical review, and promotion. The Internet is so important to successful film and video promotion and distribution that this subject is covered in detail later, in Chapter 27.

Organizations/Groups

Depending on the project, organizations and groups can present a lucrative market. On the local side, Rotary, Kiwanis, Jaycees, and other community-based organizations are always looking for ways to inform and entertain their membership at meetings and events. What better attraction than screening a locally produced documentary concerning an issue or aspect of the community?

There are literally hundreds of these opportunities available, and a quick glance in the phone book or online will highlight the numerous athletic, business, fraternal, human service, labor, political, religious, senior, veteran, and youth organizations meeting regularly within your town or metropolitan area. Buyers in this category are easy to locate (the president of the club or association) and have access to funds, advertising, and connections for speakers and guests.

Nationally, the market is enormous, as groups and organizations exist for every interest, hobby, and belief imaginable. Most provide some manner of offering merchandise to their membership (i.e., newsletters, Web sites), and sales of a film or video applicable to the group's passion is a natural. Renting a booth at annual trade shows and appearing at gatherings of the largest organizations offer another sales opportunity for independent film and videomakers.

The programming boards of schools and universities also represent a strong market for independent filmmakers. Like groups and organizations, local and national potential exists. A 30-minute math instructional video could find success with school systems across the country, while an hour-long look at the historical highlights of your city should be stocked by every school library in town.

Colleges are constantly trying to balance student entertainment demands with limited budgets. That's why the purchase of independent movies, for both library use and projection, hold such appeal for this group. The best part is, pretty much anything goes, especially if you are willing to make an appearance with your flick. Buyers in this category are usually easy to access and very enthusiastic student association leaders, making this sales option a good place to get your feet wet before approaching more sophisticated markets.

A special case should be made for the Video Software Dealers of America (VSDA), their regional associations, and other industry organizations such as local or regional film groups. Many local VSDA chapters have buying groups that make quantity discount purchases for members. This is an ideal audience, as it allows you to target a group of stores with a single sales effort. The same goes for filmmaking clubs and societies. They sponsor workshops, seminars, and events that usually feature working professionals as speakers and guests. Members are obviously film fanatics, and are more than interested in purchasing independently produced product if it is priced right.

Libraries
Personally, some of our most profitable sales have been through library buyers. Usually operating in groups of several local branches, libraries will purchase a copy or two of your production for every location in the system. For example, we sold *Pig* to more than 26 libraries in a single sale. Multiply that by the enormous number of library systems within a single state alone, and you start to understand why this is such a prime opportunity. Part of the reason a library exists is to expose communities to new forms of art, education, and self-expression, and you'll find these buyers well-educated people responsive to the plight of independent film and videomakers. Plus, they get a big kick out of talking to the actual creator of a project, making for a very pleasant sales experience (which, unfortunately, isn't all that common).

Strategic and Potential Partner Companies
Suppose you made an instructional video on professional car detailing techniques. Who are your buyers? Auto Zone and Napa retail stores would be a good start, as would any car enthusiast organizations. You

might also want to shop it to some of the many automotive-oriented cable channels and Web sites for broadcast and streaming options. And don't forget special interest distributors who specialize in sales of instructional tapes, and all of the libraries housing large do-it-yourself sections. Not too shabby of a marketplace for an independently produced video made in your garage.

But more opportunities exist in the form of strategic and potential partner companies. How about soliciting local new and used car dealers? They could offer the tape free of charge to every customer buying a new vehicle. Did you consider the companies that manufacture the detailing products? I don't know about you, but I think it'd be pretty cool to get a free video packaged with my next can of car wax. If your video is priced appropriately, these kind of value-added products hold much appeal for companies looking for an edge in a highly competitive market. Be aware, however, that partnering with large national or global organizations will take time.

These promotional opportunities are perfect examples of why you should start to market your video the day you decide to make it. Obviously a manufacturer would want his product featured in any kind of give-away video. By arranging this kind of sale before production starts, you'll be in a much better position to customize the tape to your buyers' specific needs.

Theatres and Other Public Venues

If you are promoting a feature film, you don't want to overlook local movie houses as a potential sales outlet. In addition to serving as the perfect location for a premiere (more on that later), showing your flick on the big screen as a special engagement or midnight movie can be profitable to both you and the theatre owner. Filling slow and unused time slots with a "novelty" product that holds strong appeal to the local audience is always a welcome ally in a theatre owner's continual battle against ever-decreasing profits. Though many movie houses are owned by national chains, a good number of small, regional ones do exist in almost every state. They are often willing to try new methods of attracting an audience.

For those interested in a national **four-walling** campaign, which means you literally rent the four walls surrounding a movie screen, many so-called "art" theatres specialize in independent film fare. Though the buyers at these establishments usually deal with distributors, they are not averse to dealing with the independent promoter out on her own.

Other public venues in which to project your film or video include bars and night clubs, coffee houses, health clubs, salons, and anywhere you can think of that features a screen or monitor continuously providing entertainment or distraction to patrons.

Other Sales and Rental Outlets
Movies and videos are not strictly rented and sold in the entertainment-specialty retail situation. There is an endless array of outlets that handle the same kind of transactions, though often overlooked by the independent producer as viable markets for his products.

As you move through a normal day, keep a log of all the places you encounter that either sell or rent videos. Some to look for include: convenience stores, drug stores, gas stations, grocery stores, health clubs, catalogs, hospital gift shops, airports and airplanes, fast food franchises, card shops, and hotels.

THOSE WHO WILL PROMOTE YOUR PRODUCT
Many of the above buyers and sellers will also work to promote your product, an action that ultimately helps their sales efforts. After all, as they say, the biggest part of a sale is the sizzle, not the steak. But in the truest sense, only a couple of audiences will strictly promote your film with no intention of making a direct profit from the activity.

Media
Newspapers, magazines, newsletters, radio, broadcast and cable television programming, and the Internet will all become part of your promotional campaign. Within each of these separate media lie opportunities for the promotion of your project in an editorial context. That means you won't pay for the publicity. Unlike media advertising, media promotion — interviews, feature stories, reviews — is free. You just have to sell the reporter, editor, or reviewer on the uniqueness of your

film or video. Long ago, just the fact that someone made a film outside of the Hollywood system was news enough itself to warrant a story or two in the nightly paper. Not so today.

I found this out the hard way through a very humbling experience. In the Northeast Ohio area, *Scene Magazine* is one of those weekly entertainment tabloids found in most major metropolitan areas. The paper runs movie, theatre, club, television, and sport's news of all kinds. I figured our production of a feature film centering on a somewhat controversial subject (the secret life of small-town cops) would be good enough to get us the cover. After a less than enthusiastic response to my initial query during the production phase, the editor told me he'd wait till we were finished before considering any coverage. With the film finished, I promptly sent off a screening copy to *Scene*, per their request. One week later, I got the screener back in the mail with a note telling me they weren't interested. Worse yet, when I pulled the tape from its package, it was obvious by looking at the video spools that the editor only watched about 15 minutes of the movie before making his decision. So much for local fame and fortune.

Luckily, this setback reminded me to always have a story to tell. That's part of the sizzle. And just like you'll need it for viewers, buyers, and sellers, you're going to need some excitement for the media as well. Just as with distributors and the Internet, working with the media is a special skill integral to a fully realized promotional plan. Techniques of all kinds are covered in detail in Chapter 18, devoted to the media.

Publicists
A good publicist will make you and your film household names. At least that's what she'll have you believe. But that may be exaggerating a little. The truth is that any publicist worth her fees will get you known to the right people — which includes anyone inside the entertainment industry who can help you succeed. A publicist is also responsible for getting you as many media placements as possible — getting you publicity. So whether that's setting up an interview with National Public Radio, getting you and your film mentioned in *Premiere's* "newcomers to watch" section, setting up screenings with influential acquisitions agents, or simply handing out flyers at a film festival in which your

movie is showing, publicists will put you on the map. Though usually working for hire, a recent trend finds some publicists getting involved with independent movies on a deferred basis, allowing the cash-strapped filmmaker access to a professional promotional machine. Appealing to a publicist's artistic side while at the same time present-ing your project's worth (either in a social or financial sense) is a good way to find yourself operating in this kind of set-up.

Festival Screeners

Supposedly, you cannot influence festival screeners with a slick promo-tional campaign. Submit your materials and entrance fees on time, per the instructions, and you have just as good a chance as anyone when it comes time to select the winning entries. Of course the quality of your film is the most important determinant of this process. But a success-ful front-end promotional campaign, one that's created a high level of awareness for your project, won't hurt your odds. It is also a good idea to establish a personal connection or two with someone on staff. Make some phone calls. Try to figure out if your project is a good fit for that specific festival. Send a note of thanks after the conversation, or some quick-to-digest promotional piece (postcard, flyer, or newsletter).

Screeners watch all the films that ultimately get accepted to or rejected by any of the hundreds of festivals held every year. By knowing the cri-teria, likes, dislikes, and quirks of a festival before entering, you'll have a better chance of ensuring your film's appeal to this group. And it doesn't hurt to have an industry-insider contact. Though they'll claim otherwise, festivals offer the single most persuadable outlet when it comes to using the influence of "connections." Once your film is accepted, festival staff will become some of your greatest allies. It is in their best interest to trumpet your work, and they'll do so vigilantly.

Chapter 3 Summary Points

- You must be able to identify an audience for your project before promoting and selling it to them.

- There are really three audiences for your film or video project: those who will watch it, those who will buy/sell it, and those who will promote it.

- By questioning your potential audience's gender, age, education, religion, race, occupation, and income, you can create a profile that will identify those people most interested in your project.

- Profiling and identifying a sizable audience can often spur the production of a film or video.

- It is sometimes discovered that films and videos can appeal to two distinct and disparate audience profiles.

- Distributors, producer reps, distributor reps, agents, home video, cable, pay-per-view, Internet, and broadcast television buyers are often more important to the success of your project than the "ultimate" viewer.

- Other non-traditional buyers and sellers of entertainment product exist, including organizations and groups, libraries, strategic and potential partner companies, theatres, and other public viewing venues.

- Members of the media, publicists, and film festival screeners will be somewhat responsible for promoting your project.

WHAT ARE YOUR HOOKS?

A good selling point for your film or video — or any other product for that matter — is known as a marketing hook. This hook is what "catches" your intended audience and, taking the fishing allegory further, "reels" them into the theatre, video store, or retail outlet to experience your project.

Hooks are exploitable elements, aspects associated with the production that will attract those people who comprise your audience profile. Hooks take many forms: a person, place, thing, action, or idea. Many times, a film or video's title alone will function as the marketing hook. Finding the right hook means a mediocre film can become wildly successful while a much more accomplished, yet hookless, project sits on a shelf.

The survival and success of most independent films and videos are predominantly dependent upon realizing the proper hook to use in the promotional campaign. Unlike blockbuster movies, whose promotional plans may be changed weekly until they hit on the supposedly correct selling point, you'll be dealing with a limited budget and probably only have one chance of finding and promoting the proper hook. So you better be on target... or at least very close.

In discovering the appropriate hook for your project, you'll need to re-examine your concept capsule and audience profile. You know what you have. You know who wants it. Now it's time to decide *why* they want it by answering the question:

What is the single most promotable element of your film or video that will attract its target audience?

To begin with, realize that, unlike Hollywood films, big-budget instructional tapes, or National Foundation of the Arts-backed documentaries, you will not have the following possible hooks to exploit:

- Critically acclaimed director
- State-of-the-art special effects
- Well-known celebrities
- Access to a national or worldwide current event
- Exotic locations

Any one of the above can and have been used on their own as a promotable element in drawing audiences to a project. Arnold Schwarzenegger movies appeal to a certain segment of the population based solely on his star appeal. Spielberg flicks do well because of his reputation as a great director. *Titanic* was sold on its amazing special effects. *Get on the Bus* found an audience because of the media exposure given to the Million Man March event on which it was based. And *Lifestyles of the Rich and Famous* is still playing after all these years because it gives the general viewing public access to locations not likely encountered on a daily basis.

Keep in mind that even though a project possesses one of the above hooks doesn't mean it will find success. Some of the more notable examples include:

- *Wild Wild West* (popular star: Will Smith; name director: Barry Sonnenfeld; amazing effects)
- *Ishtar* (popular stars: Warren Beatty, Dustin Hoffman; beautiful locations)
- *The Last Action Hero* (popular star: Schwarzenegger; name director: John McTiernan; multi-million-dollar effects budget)
- *Battlefield Earth* (John Travolta; state-of-the-art effects)
- *Book of Shadows: Blair Witch II* (how could this fail?)

HIGH CONCEPT

All the more reason that you will need to find what the entertainment industry likes to call **high concept** to serve as the hook for your project. What this means is that you must reduce the elements in your film or video into a promotable idea that can be conveyed through many mediums — from advertising and public relations to sales letters and key art. *The Blair Witch Project* is a textbook case of high concept. A low-budget film shot on videotape, the high concept idea presents a group of researchers disappearing with only a videotape record of their activities left behind. It's no secret that a $7 million marketing campaign helped sell this concept to the masses. But it was the idea — not the stars, budget, locale, or special effects — that captured audience interest. And it's that kind of idea you'll want to create to promote your project.

How do you find that element, the so-called high concept, that will make an audience want to see your film or purchase your video over other movies and tapes? It comes down to analyzing what you have and distilling those promotional elements that touch upon an audience's emotions into an attractive, cohesive, and sensible concept.

ANALYZING THE ELEMENTS

Whether your film or video is still in the planning stages or already "in the can," you can get a good start on this process by examining the following:

Story – Can the story of your film be the main attraction? Go back to your concept capsule. Are there emotions present that "talk" to the potential viewer? Can audience members identify with the feelings expressed through the action of the story? Is the script extremely well-written, featuring great irony, insight, or plot twists? Does the story deal with survival, love, hate, or fear? Hanging a promotional hook on one of these four basic emotions can work well if the story truly delivers. *Cast Away* featured amazing special effects and Tom Hanks. But the reason viewers flocked to the movie was its incredible tale of survival and love. The same goes for *The Sixth Sense*, where the movie's premise was based on a single and very unexpected plot twist.

Source – What is the source of the story? Is your project based on a famous novel or the work of a renowned author or journalist? Does it depict a current event, true-life drama, or historical account? The words "based on a true event" and "based on the writings of" can become pretty powerful promotional hooks. An all-star cast and state-of-the-art effects certainly didn't hurt box-office potential for *The Perfect Storm*. But part of the promotional effort was aimed at viewers familiar with the source of the story: an article in *Men's Health* magazine. Don't believe the work of mainstream authors is out of your price range as an independent, either. Joyce Carol Oates and H.P. Lovecraft are two examples of famous writers agreeing to low-budget, independent adaptations of their stories.

Genre – Does the genre of the film or video contain its own hooks? Will audience members want to see your project just because it belongs in a category of movies they enjoy?

- Martial arts and action fans will want a basic good vs. evil set-up. Present the story with a different twist (familiar situation, different concept) and you'll garner even more attention. Name stars usually provide the hook for many action-oriented flicks, but this segment of the market is full of independent success stories.
- Westerns need authentic locations, costumes, and dialogue to draw loyal audiences.
- Horror movies rarely fall out of popularity. Slasher films dominated the early 80s, and the genre rose again in the late 90s with the likes of *Scream* and *I Know What You Did Last Summer*. On video, horror movies continue to score as top renters.
- The audience is growing for science fiction, occult, and unexplained phenomenon projects. Productions dealing with the subject matter do well in consumer direct sales and on cable television.
- Comedy is tough. If it is hilarious, or delivers a story in an offbeat manner like *There's Something About Mary*, audiences are guaranteed. Anything less usually falls short. Comedy is sold on its stars, though no-name laugh fests can find an audience via video and PPV (Pay-Per-View).
- Documentaries attract audiences based on their subject matter. If you are a fan of the rock 'n' roll scene, then chances are good you've watched *Gimme Shelter*. However, that doesn't mean you've also seen *The War Room*.
- The hook of an instructional video is that the audience member is seeking that kind of specific knowledge. From deck-building and house painting to public speaking and

crocheting, if your projected viewer base is large enough and accessible, and your budget is appropriate, videos on these and hundreds of other topics can find success based solely on the instruction presented.

- Reality-themed projects are low-budget sellers that usually find their audiences. Whether it's *Good Animals Gone Bad*, *The Girls of Spring Break Daytona*, *Survivor*, or *The Blair Witch Project*, viewers love to watch other people in embarrassing, fearful, or difficult situations.

- Independent or art films have become their own genre. Beginning with the likes of *Clerks* and *Slacker*, a movie that fits this category is hard to promote, as often its only hook is that it was completed on a low-budget outside of the Hollywood system. Once upon a time this was a selling point, but the concept has been done to death. The proliferation of independent films is astounding. With digital video editing equipment readily available, anyone can become a filmmaker — and has. Just ask any distributor whose desk is flooded with screeners from hopeful producers. Don't make the mistake of hanging your promotional campaign on the fact that you and a bunch of college pals wrote, directed, and produced a film while waiting for the pizza delivery man every night. It's not news, so it won't work as a promotional hook.

Celebrity – Does your project feature a celebrity of some kind? Obviously you won't be working with "A" list talent out of Hollywood, but maybe you were able to rustle up a sports hero, on-air personality, cult figure, or faded genre star. *Killer Nerd* featured Toby Radloff, an MTV reporter familiar for his claims of being a true-life nerd. That the film featured a so-called genuine nerd — a person also popular with a national youth audience — acting in a movie about nerds was one of its hooks. Buster Crabbe, a movie idol from the 40s and 50s, was featured in Fred Olen Ray's ultra low-budget *The Alien Dead*. By no means a starring vehicle for the aging actor, the producer used Crabbe's "marquee value" to help sell the film to the public. Many of the *Survivor* stars have become the promotional hooks for films, television specials, and videos featuring their talents. If your project can, or does, feature a public figure of any kind, investigate exploiting that angle as a marketing hook. Even on a very local level, it works. Maybe you are planning a video on the history of your town. Your market of libraries, schools, local retailers, and video rental shops could broaden to include local television and cable stations if the project features a political or social personality well known in the surrounding community.

Creative Light Entertainment's Suckers *features two kinds of hooks. The first appeals to the common experience of car shopping (the movie's tagline is "Whatever You Do, Don't Buy Your Next Car Without Seeing This Movie"), while a recognizable television actor, Daniel Benzali, is brought in for the second: star power.*

Special Effects – Are special effects the star of the movie or video? While *Friday the 13th* presented nothing new with regards to its revenge-themed storyline, the never-before-seen gruesome make-up effects brought in theatregoers by the millions to the independent, low-budget thriller. The ability to expose a viewer to something that doesn't exist in reality is a great hook. *Star Wars* offers a traditional good vs. evil structure. But the special effects can be credited for a large part of the movie's initial success. Advances in computer animation have led to projects like *Final Fantasy: The Spirits Within* and *Shrek*, both audience favorites promoted in some part on the basis of their ground-breaking special effects.

Location – Does the film or video feature unusual locations? Did you shoot in a volcano, cave, genuine haunted house, or underwater? Who doesn't enjoy the *Nature* series on PBS? The audience is guaranteed to experience an unexplored (to them) environment each episode. Mary Kate and Ashley's latest video probably doesn't need a hook beyond the popular twins to make sales. Yet featured prominently on the box of their *Holiday in the Sun* is the claim "Filmed in the Bahamas."

Uniqueness – Does the film feature something that is "one-of-a-kind"? Can you offer the audience a chance to see something that they can't see anywhere else? Are there bizarre or unbelievable elements attached? Probably one of the best examples of using this kind of hook: *The Alien Autopsy* program, originally broadcast on Fox, went on to sell hundreds of thousands of copies on video.

Exploitative – Is the film or video exploitative in nature? Similar to the above, in that the viewer is promised access to usually inaccessible subject matter, exploitation projects often take advantage of the audience through cheap thrills. *The Faces of Death* video series is legendary for its ineptitude and fakes. The tapes still made millions even though they rarely delivered the goods. Exploitation movies and videos can take many forms — from feature films to 30-minute cablecasts — and often deal with tawdry subjects. If the whole point of your movie is to exploit an audience, you'll probably find one, but be warned that careers taking this direction are usually short.

Title – Can the title of your project work as your hook? High concept ideas must be carried over into the film's title. This word, or group of words, is your most visible chance to present the audience with the hook about the project's content. *Pig* started out as *Small Town Cop*. While the latter title definitely tells the story of the film, *Pig* is a much more effective hook. It suggests a definite idea in most people's minds about the character and the subject of the film. Think about what the words in your title mean and what message they communicate. Also consider the emotions evoked from potential viewers reading your title.

Price – Can the price of your project be used as a hook? Though pricing theories will be explored in Chapter 22, you may want to think about this option now if your film or video appears to need an extra promotional push. Possibly you've created a self-defense instructional video that really doesn't have any distinguishing characteristics. The tape presents solid information in an easy-to-understand manner, but doesn't feature any exploitable elements, exotic locations, special effects, celebrities, or story line. After researching the market, you find that most self-defense instructional titles sell for $19.95 and up, which leads to your price hook: "Learn to defend yourself for $10." That's a pretty tempting offer.

A hook must contain elements specifically attractive to your targeted viewing public, while at the same time setting you apart from similar product in the market. To get the proper perspective on this dilemma, put yourself in the shoes of someone in your target market. Let's say you've made a horror film that is going direct-to-video. Ask yourself, "If I'm shopping for a Friday night rental, why would I pick my film over all the others around it?" Obviously promotional art is a selling point, but what other values does the production possess that make it appeal to the intended viewer? What element can be promoted to make your film required viewing? Is it scarier, gorier, something that could happen to anyone, or does it just happen to have the best title on the shelf?

Maybe you are the proud producer of an independent art film. You know, one of those movies featuring five or six twenty-somethings discussing the philosophy of relationships, work, and beer while the camera rolls. How will you promote this kind of project? Is it the low-budget version of *The Big Chill*? If so, study the way that movie was marketed, specifically looking for its hooks. Maybe you decide it's not your film's great conversation and insight that will reel in an audience, but the riotous jokes and scenes of physical hilarity that break up the dialogue. So what was once a semi-serious drama on growing up is now targeted at the beer-pizza-and-a-movie party crowd. Continually redefining what you believe is your project's most marketable and targeted concept is the key to developing your promotional hook.

Recognizing a high concept hook is not something that happens quickly. To get in the practice of identifying successful approaches, try studying the past year's most popular movies. Most blockbusters do more than offer eye-popping special effects or an all-star review. Those elements are probably part of the package, but rarely stand alone. A unique storyline, something the audience remembers and enthusiastically finds interesting, is usually the source of success for most of the big money-making films.

When developing a hook, consider all audiences buying the product. Remember, there are more folks in the sales pipeline than just the ultimate viewer. Retail stores, distributors, television, cable and pay-per-view stations, film festivals, and others demand the same type of "courting" as your viewing audience. While a high concept hook is universally important, each of these groups have their own distinct set of buying characteristics and you may have to devise a secondary hook for each market solicited. In this situation, the secondary hook is more of a sales technique, such as packaging, pricing, and delivery terms, rather than an idea or concept that describes your film or video.

Finally, keep in mind that the development of a marketing hook (as well as other promotional practices) will more than likely be handled by someone other than yourself if you secure a distribution contract. This is not to say that it's a waste of time to work through the above exercise and find the core attractive quality of your film or video. Analyzing your project in such a way can only lead to a better understanding of what you've done and what you need. And, a well-developed promotional campaign — one that utilizes a carefully considered high-concept hook — may actually help in the chase of the fabled distribution contract by presenting the buyer with concrete ideas on how to attract a sizeable and profitable audience segment. But any respectable and experienced distributor will use the services of an in-house or for-hire marketing department skilled in the execution of such campaigns. From the title of the project to the key art to possibly even identifying a new target audience, professional marketing personnel may change the whole approach to securing sales and success.

Before we start chasing distributors, you need to investigate what is really motivating you, what you ultimately want to achieve through the realization of a promotional campaign.

Chapter 4 Summary Points

- A project hook is an exploitable element associated with the production that will attract those people described in the audience profile.

- An independent film or videomaker needs to discover the single most promotable element of his project before designing a promotional campaign.

- A project's story, source, genre, cast, production values, location, uniqueness, title, and even price can be used as a hook.

- Though a hook must "feel" familiar to an identifiable audience segment, it must also set your project apart from all other offerings in the market.

- Consider all audiences, not simply the final viewer, when determining a marketing hook.

WHAT IS YOUR GOAL?

Without the proper information, defining an accurate goal for the outcome of your project's promotional campaign can be tough. What do you use as a parameter to measure your success? Performance of similar films or videos? Income generated? Total number of units sold? Personal satisfaction? Positive reviews? Amount of media coverage?

In addition to identifying this sort of concrete criteria, consideration of less tangible information like your emotional, physical, and financial state must be part of the goal-setting process. Settling for nothing less than a national theatrical release after a particularly productive day of screenwriting is not wise. Nor is only hoping to breakeven on expenses after deciding to cast your cousin Alvin in lieu of an experienced actor who's too expensive. As these examples illustrate, it's impossible to make a reasonable decision about your project while feeling too high or too low. This is not to say that you shouldn't "reach for the stars," but:

You need to be realistic in defining your goal.

THE BIG PICTURE
Let's step back for a moment and look at the "big picture." When you began your film or video, what was your motivation for creating it in the first place? Was it merely an artistic expression, were you acting on business instincts and filling a void in the market, or did you dream of the fame and fortune that accompanies success in the entertainment industry?

Whatever the case, consider what it is you want the promotion of your film or video to ultimately attain, accomplish, or achieve, and set that as your goal. By doing so, you'll give your energy a specific focus, enabling you to better sort out what is important and what is irrelevant. By creating an achievable goal, you can move beyond any self-doubts, overcome your fears, and think "successful," which leads directly to being successful.

That all sounds great, but we're still left with the original question: What is your goal? For film and videomakers, many outcomes are possible and appropriate. Finding the goal that applies specifically to your project — while at the same time maintaining relevance to your personal, artistic, and career motivations — is a process that shouldn't be rushed. While answering your needs on many levels, this goal will also serve as a guide in determining the level of effort and types of techniques used in implementing your promotional plan. This means the amount of money, time, and sweat you expend is directly related to the goals you want to achieve. So were talking pretty significant stuff here!

MULTIPLE GOALS

It's important to note that you probably have more than a single goal. Don't worry — that's just the nature of the process. For example, you may want to capture a 10% video store penetration rate for your video (meaning, one out of every 10 video stores will stock your film on their shelf). At the same time, a 20% profit goal may exist for each tape sold. Multiple goals are valuable in that they direct your efforts for each task within the overall promotional campaign. A profit goal will help with pricing issues just as a penetration rate goal will determine the amount of video stores you solicit for sales.

TEN QUESTIONS

Before you set any specific goals, examine the major motivations driving you toward becoming a film or video promoter in the first place. To do that, take some time and honestly answer the following questions:

1. Do you want to become rich and famous?
   Who doesn't? Taking lunches with superstars, doing interviews, and shopping on Rodeo Drive doesn't sound too shabby. If that is what you want, there are probably faster routes to getting there than promoting and selling an independent film or video. Not that it's unachievable. The guys from T*he Blair Witch Project*, *Pi*, and *El Mariachi* were all catapulted to fame and fortune following the promotion of their first features. It's just not the easiest step for a camcorder-toting boy from Kansas to take. If you decide rich-and-famous is how you would like to eventually describe yourself, consider the varying

levels of this goal. By famous, do you mean in your town, to the independent filmmaking community, within a special interest group, or nationwide? The same could be said for being rich. It's a relative state, really. Wealthy to one person is vastly different to others. Having the luxury to quit your day job may be enough to make you feel rich.

2. Do you want exposure?

If so, how much, where, and what kind? Consider local, regional, national, and worldwide markets. General, entertainment, and industry-specific media. An interview in *MovieMaker Magazine* is great for a filmmaker looking to promote his name and project to the independent filmmaking community. But if your video deals with youth violence, would a story in an education journal positioning you as an expert better serve your desires? Who you want to expose is an important fact to know — one that will direct many of your promotional decisions.

3. Do you need to make contacts?

Maybe you need to find new talent, crew, or subject matter for your current or future project. Networking with fellow film and video artists, distributors, and others in the industry is easily accomplished while you work as a promoter. As clichéd as it sounds — *who you know* really makes a difference when you are working in this business. I can't tell you how many times dropping a name or having an introduction opened a previously closed door. Rubbing shoulders with working professionals is educational, fun, and often plays to the I-want-to-be-famous motivation. Decide if your desire to meet and connect with people is based on the ultimate advancement of your project or just a personal need.

4. Are you seeking an immediate, non-monetary payoff?

This is the call for many activists. A lake repeatedly contaminated by a dominating local employer is the subject of your latest documentary. The chance of your video making a change in the way things are done — in this case, stopping pollution — may be the driving force behind your work in promoting it. Finding and influencing the correct audience, whether it's a popular media reporter or simply a factory owner fearing public reprisal, is what motivates you — not Hollywood connections or a big bank account.

5. Do you want money and/or connections for your second project?
Many independent filmmakers are banking on a career in the field, with their first film serving as a resume. Do you want to gain acceptance as a working professional so that you can continue to do this kind of work? Do you have a problem with sacrificing your personal life? Do you like making phone calls, writing letters, and sending packages? Do you like dealing with rude people? Do you hate rejection? Is there nothing on this Earth you'd rather do than work on a film or video?

6. Are you selling yourself or the product?
Independent film and videomakers are often more interested in selling themselves than the product. Some successfully do both. Kevin Smith of *Clerks* fame is a good example. Do you want that elusive fame, or do you really want your project to succeed? Are you humble enough to allow others attached to your project to grab the spotlight if that's what it takes to sell it? Would you rather have your name or the title of your project appear in a newspaper article?

7. Do you need to pay back investors or yourself?
Avoiding someone to whom you owe money can be a strong motivator. If you've borrowed the lion's share of your budget from friends and family, paying them back may be your first priority. It's uncomfortable to continually defend your project's status to creditors, and hopefully you explained the risks involved before taking any contributions. Even so, many lenders will badger you about their investment and the only way to shut them up is to return their funds. Personal funding issues can be even more troubling. Bankrolling a project with your savings, credit cards, or spare cash can create a desperate situation when it comes time to objectively consider offers from distributors, especially if you've been living on peanut butter and

ramen soup throughout the entire production. The need to get "back on your feet" is sometimes an incentive that film and videomakers refuse to identify.

8. Do you just want an audience?
   The sheer joy of entertaining, informing, or affecting an audience motivates many artists. In these cases, it's not the potential riches or connections brought about by a project's success that matters to you, but the fact that someone else is watching what you created. Many one-shot filmmakers find this to be true. After all the writing, shooting, editing, and selling is done, and two years of their life has vanished, the filmmaker realizes that they were only looking for an audience — deciding there are certainly quicker and easier routes to finding one. To others, the power attached to group influence is strong enough to keep them working on new films and videos.

9. Do you believe that you are the only person capable of the job?
   Are you unable or unwilling to put your project in the hands of anyone else? Most independent films and many videos can take up to two years or more to complete. If you've lived with a project for this amount of time, it's tough trusting someone with your "baby."

10. Do you want closure?
    Like any other large-scale project, an independent film or video has a life cycle that must end. Living with all the minutia involved in such a process also demands letting go. Closure is an important goal for those who need to move on.

GOAL CRITERIA

Answering the preceding questions truthfully is just the first step in goal setting. Re-examine your answers and test each against the following criteria:

- Do your goals consider both long- and short-term consequences?
  A good goal will provide you with a long-term vision of what you want to do in your lifetime as well as short-term motivation to

keep you going on a day-to-day basis. The desire to own your own production studio is appealing. Without smaller, almost daily accomplishments (making five phone calls a day) however, the road to reaching this dream may seem too long to travel.

- Are your goals realistic, achievable, and challenging?
You must be able to balance the need to provide yourself with challenges against the need for success. Striving to sell your video to every rental store in the country may not be necessary for success. It's a great goal, but one that could lead to your biggest enemy: burnout. It is natural to build on past success to meet new challenges.

- Are your goals flexible?
You will want to revise your goals as your life, priorities, and opportunities change. Perhaps halfway through the promotional campaign, you receive a buyout offer. A distributor is willing to purchase all rights to your movie for a very attractive price. In such a situation, short- and long-term sales and media exposure goals will likely change.

- Are goals measurable?
Goals must define precisely want you want to do — with amounts and with deadlines. Say you've decided that you want to be rich. Well, what is rich? And how will you know when you've got there if you don't have some way to measure your progress. That's why it's necessary to set quantifiable goals — yardsticks, if you will — that allow you to gauge your efforts against a predetermined anticipated result. A goal of selling 20 videotapes per week, for the next eight weeks, at $20 each, is a quantifiable goal.

- Are they *your* goals?
Be sure goals reflect your desires, your needs, your aspirations. Proper goals must reflect the manner in which you like to operate. You cannot successfully reach a goal that forces you to work against your nature. Setting a goal to make five

personal sales calls weekly is not going to work if you are shy about meeting and talking to new people. Revising the goal to make five sales approaches (via letter, e-mail, or fax) is more attainable and in tune with how you work.

■ Are your goals in writing?
Finally, put your dreams and aspirations in writing. Post them where you must see them at least once a day (the refrigerator is a good spot). This makes them real hard to ignore or forget.

Chapter 5 Summary Points

- Goals set for the outcome of your film or video's promotional campaign must be realistic.

- Goals are necessary to give your efforts focus.

- It is acceptable to have multiple goals.

- Examine personal and professional motivations when setting goals.

- Goals must be flexible, measurable, and applicable to long- and short-term situations.

- Be sure any set goals are *your* goals.

A MEDITATION ON LIMITATIONS

I have waited until now to discuss limitations for one important reason:

If you are told what you cannot do before you begin to do it, you will be severely restricted in what you aspire to do.

In other words, if I presented the limitations facing you before you created a concept capsule, audience profile, hook, and goals, I doubt if your aspirations, dreams, ideas, and plans would be as big as they are right now. You need to be able to think freely about your project, with no restrictions placed on your creative abilities. You also need to look at the limitations facing you (and every other independent film and videomaker), and realize they are surmountable if recognized and dealt with.

INADEQUATE FUNDING

Inadequate funding is a common denominator among independent producers. You wouldn't be categorized as such if you had $10 million to blow. Money was scarce enough to find during production, much less having excess lying around for a large-scale promotional plan. This thought is the number one excuse many one-shot, low-budget filmmakers use for letting their films languish on a shelf. They mistakenly think that without a large amount of money, or the services of a recognized distributor, they cannot properly promote their film. These people end up believing filmmaking is a waste of time and money, and go on with their lives, never to venture into this wonderful industry again. Don't worry. You will not become a member of this group.

Many methods or promotional techniques in this book will cost you little more than the price of paper, phone calls, postage, and some hard work. If you believe in your film's success, as I'm sure you do, a shortage of cash won't stop you. People are impressed more with your relentless ambition to succeed than they are with the ability to plunk down a wad of cash. It is this fact that will open promotional doors for you.

Hopefully you're feeling better about your low-cash situation. However, your project's still lacking stars, good acting, exotic locations, special effects, and top-notch cinematography. No doubt about it, the odds are stacked against you. But there's a way out and around this situation. There always is — as long as the project has merit.

DOES IT HAVE MERIT?

Some projects, unfortunately, are just bad. If you find yourself the owner of such a film or video, you may want to chalk it up as an expensive and timely learning experience, and quietly place it on the shelf. There are untold numbers of Hollywood-produced films that fall to such conclusion. Deemed unreleasable, the film is shelved rather than embarrass the talent or producing company. Tricky marketing stunts and gimmicks can sometimes save the situation. York Home Video released the Carmen Electra film *American Vampire* with dubbed-over dialogue to hide the original tracks. Supposedly the movie was so bad, they thought spoofing it a la Woody Allen's *What's Up, Tiger Lily?* would salvage some of its value. The tactic worked to an extent, selling a respectable number of video and DVD copies to video stores nationwide. You can't, nor do you want to, attempt this kind of promotional chicanery, however, as you will quickly earn a reputation for producing and releasing shoddy work, and lose all chances of developing loyal audiences.

DON'T FORGET YOUR CREATIVITY

Projects that do hold promise, and deliver quality entertainment, instruction, or information, can and should find their audiences if promoted properly. Creativity is often the key. Anyone who takes it upon himself to produce a movie is creative, and should use that skill to his advantage.

Think your film doesn't have any promotable elements? Maybe location can become your hook. Let's say the shoot went down in Grange, Idaho, a town offering little more than potatoes and power lines. Promote it as the first film ever to be produced in that part of the country. Still not enough heat to draw media attention? Do any urban myths exist about the locale — possibly you shot over a former cemetery? Tell the media about ornery apparitions visiting the set. Watch how quickly reporters start paying attention. Remember when the Red Hot

Chili Peppers recorded *Blood Sugar Sex Magic* in a supposedly haunted house? Sure the CD rocks, but the hook of that spook-filled mansion garnered tons of publicity and fan attention.

The independent video feature Vixen Highway *was promoted about as far away from Hollywood as one could get — the Minnesota tundra. That fact actually helped John Ervin and his crew when it came time to promote the flick.*

Continually ask yourself what aspect of your film makes it different from all others. Remember, people will want to see something if they think they are unable to see it anywhere else (the whole reason behind the success of circus sideshows). Go back through the hook questions and try to pinpoint what feature of your film gives it its unique "oneness." Find that angle and work from there. Don't give up.

YOU ARE NOT WORKING IN A VACUUM

As much as you'd like to believe in your originality, chances are good your idea, script, movie, or marketing ploy is not original. Pretty much *everything* has been done before. Sure, there are exceptions — but those are rare and usually it's the execution that is different, not the original idea. Filmmaking is an *extremely* competitive and busy business. You already know that thousands of independent films and videos are produced annually. And the limited numbers of outlets for these productions get inundated with submissions from film and videomakers all over the world sharing the same dream as you. Most wide-interest cable networks (Discovery, History Channel, Sci-Fi Network) will receive upwards of 2,000 independent submissions yearly. Mainline distributors regularly sift through hundreds of solicitations a month. And you know the odds of getting a film into the Sundance (or even Slamdance) festival. Regardless of personal feelings, your movie or video is just one of a hundred or more possibly very similar submissions in any one acquisition cycle. It's often not "special" to anyone but you. That's why you have to promote the hell out of it.

SAY GOODBYE TO PERSONAL TIME

A final limitation to consider is the amount of personal time you surrender to support the marketing effort of your picture. Are you willing to stay up late, get up early, write hundreds of letters, and constantly call reporters, distributors, and retailers just to get your film promoted? I hope so. The amount of time you dedicate to promoting your film is directly related to the amount of promotion it receives. I realize you may have a real job, a family, a home, and other aspects of your life that demand your attention. But you only have one chance to promote your film in the right manner before time passes and the public is interested in something else. I can't even guess at the amount of time this process may take. Sometimes you only send one letter to get the job done and other times you may be required to place five or six follow-up calls and resubmit the original information... twice! The thoroughness of your campaign and the range of exposure you receive are up to you. If you're only interested in receiving local attention, the job isn't too tough. But if you're considering national media awareness, prepare yourself for some late nights.

Limitations to your success in promoting, selling, and distributing your film are only as large as you perceive them to be. If you're willing to take some time to work at it and your desire for success is great enough, there is no limitation that you cannot overcome. With that sermon thankfully over, let's move on.

Chapter 6 Summary Points

- Limitations should not restrict your creativity or desire to succeed.

- Inadequate funding is not always a deterrent to a film or video's promotional campaign.

- A project lacking merit is always a tough sell, and if you manage to fool an audience into buying it through deceptive marketing ploys, you are jeopardizing your future integrity in the industry.

- Creativity can overcome many barriers to success.

- When promoting a film or video project, do not ignore the demands it places on your personal time.

HOW DO YOU DO IT?

PART ONE:
FINDING YOUR AUDIENCE,
DISTRIBUTION, AND
SALES THROUGH A
DISTRIBUTOR

While there are numerous avenues for every film and video in reaching its audience, the most immediate link to viewers is through a distributor. The veritable "holy grail" of independent film and videomakers, distributors offer you the best chance of widespread success with your project, and have earned this reputation by:

- Being well-connected – They know folks up and down the entertainment industry ladder.
- Having the ability to sell in many markets – Distributors specialize in promoting and selling product regionally, nationally, and globally.
- Knowing what sells – Through experience with different markets and products, a distributor can often accurately anticipate what will be popular.
- Knowing when to sell – Distributors not only predict market trends, but many times create profitable selling environments through price negotiations and release patterns.
- Knowing what to charge – Distributors are keenly aware of what the market will bear for any specific product.
- Knowing whom to solicit – Distributors won't waste yours, a buyer's, or their own time by approaching the wrong people.

And, maybe most importantly, distributors add clout to your project — an intangible value that can be priceless when it comes time to put on your promoting shoes. Distributors will sell your project to buyers you never could have reached on your own, or never would have known existed. It is this high level of connections, experience, and expertise that makes working with a distributor so valuable to the independent producer.

In strictest terms, a distributor is a major entertainment industry conglomerate, such as Universal or Warner Brothers, which will distribute your film or video to all audiences via every conceivable media throughout the world. A specific release hierarchy is followed when distributing a film or video to ensure maximum profitability from each entertainment marketplace. The product is generally exploited theatrically first, then on video and DVD, followed by television. Once the product reaches the television market, the sales order is pay-per-view, pay cable, network broadcast television, basic cable, and broadcast syndication.

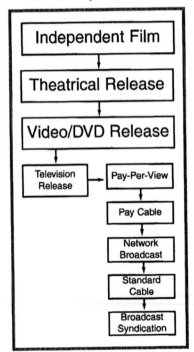

The hierarchy of an entertainment product's ordered selling

ORDERED SELLING

The reason for a distributor's "ordered" selling is that a theatrical release offers the best opportunity for recoupment on a project's cost. A $30 million movie, unless specifically designed for small children or some other home-based market, demands to play in theatres. Whether profitable during its theatrical run or not, a release of this kind creates a huge amount of publicity and public awareness, which, in turn, will drive sales in the video, cable, and other ancillary markets. The films consumers choose first in PPV and video rental situations are those they learned about from the advertising and publicity surrounding their initial theatrical release. Obviously, the smaller the budget, the less important theatrical becomes (except for awareness), but it is the primary market for distribution in many cases.

Another reason for ordered sales, regardless of whether a project begins its distribution process theatrically or not, is to avoid audience and demand conflicts that would arise if a product was released concurrently through different media (television, theatre, home video). If a movie is available on video at the same time it plays in theatres, viewers will undoubtedly wonder about its "newness." Except for the biggest blockbusters (which at times are re-released theatrically after being available on video for years), product desirability always suffers in this scenario.

At the same time, selling "out of order" forces you to cannibalize high-profit sales with lower-profit deals. Imagine your independent film has just played on a basic cable station. Do you think you'll be able to sell it to a PPV service now? Neither the service provider nor the viewer would want to pay for something that previously was offered for free. However, film sale situations don't take place in an ideal environment. What if the basic cable station was offering a hefty sum for the program and no solid PPV deal yet existed? Do you take the sure money and squash any chance for bigger market sales? Or, do you wait out a potential second deal, possibly killing the first offer? These are the kind of decisions distributors wrestle with daily; the industry runs on their input and experience in such matters.

A quick note: To an extent, sales to foreign, institutional, and educational markets can take place at any time throughout the distribution cycle, as these deals will not usually affect the product's attractiveness to U.S.-based buyers or viewers.

PROLIFERATION OF COMPANIES

Years ago, a single distributor was able to shepherd a film through its various sales opportunities. Many foreign markets and modern media such as the Internet and home video didn't exist. Now the world, and the number of media outlets available, have grown too large and numerous, making this task too costly and complex for a single company to handle alone. And that's where things get confusing. Today's distributors often license and sell many of a product's ancillary rights to other companies, allowing for the greatest realization of the film or video's market potential. For example, Miramax may acquire your independent feature for festival and theatrical release, license its domestic (meaning within the U.S.) video/DVD rights to a specialty video distribution company, sell its broadcast rights to a firm with expertise in network television deals, while holding its foreign, pay-per-view, and Internet rights for other companies that handle those markets. This process can take place with every market in every country, creating a maze-like path for the product to follow in reaching an audience.

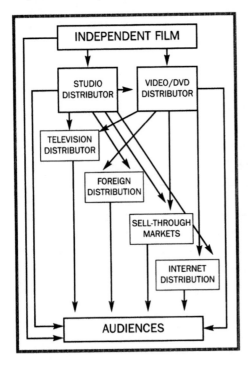

The maze-like path an entertainment product can follow in reaching its ultimate audiences.

To add to this confusion, there are over 500 companies calling themselves "distributors." This can include traditional distributors, rack-jobbers, sales agents, wholesalers, and leasing companies with selling territories around the globe. Further confounding the process is the fact that many executives working at these distribution companies quit, lose, or change their jobs frequently, making it nearly impossible to keep track of reliable contacts. And more than a couple of corrupt operators are always looking to make a fast buck off of your hard work. Considering all of the above, it becomes quickly apparent that the search for a qualified distributor, one well-suited for the subject and scope of your film or video, is not going to be easy. But it's not impossible. You just need some education on how the distribution companies operate and interact with each other, the audience, and product suppliers like yourself.

TYPES OF DISTRIBUTORS

Distributors can be categorized in many ways — size, ownership, operations, markets, and product line are just a few of the more popular defining criteria. For purposes of this book we'll examine distributors in the following categories:

- Studio

- Independent
- Sub
- Wholesale

Many more types and levels of distributors exist to service the flow of product from production to viewer. But the distributors that fall within these groupings are those companies offering both immediate access to their services and the best opportunity of securing a release agreement for your independent film or video project.

Studio Distributors

As the name implies, these are the big boys — Warners, MGM, Columbia TriStar, Universal, and the like. The movers and shakers in the industry, studio-based distributors exist to primarily release their own product — that is, distribute the films made within their own studios (though many cross-studio production and release deals have been made in the past several years, i.e., a movie made by Paramount may get picked up for distribution by Warners). These companies also actively acquire independently produced films for distribution under their company banner, as it would be cost- and time-prohibitive to create in-house the total amount of product needed to fill annual release schedules.

Depending on whom you are talking to, there can be anywhere from 10-25 companies included in the "studio" group of distributors. The more obvious ones include:

- Buena Vista (Disney)
- Universal
- Warners
- MGM
- Paramount
- New Line
- Miramax
- Fox
- DreamWorks
- Artisan
- Columbia TriStar

The reason the list varies is because many of the above distributors are actually owned by or partnered with another company on the list. Buena Vista owns Miramax. Warners and MGM are linked, and so on.

Studio distributors sometimes function as direct distributors. This is most common in theatrical markets where the large-scale movies produced under the various companies' trademark guarantee screen positions at multiplexes across the nation. Universal distributed 2001s *The Mummy Returns* to thousands of movie screens upon its release. The highly anticipated sequel was both made and directly distributed theatrically by the same company.

Universal also released *Tremors 3* in 2001. The flick skipped a theatrical release and went straight to the video/DVD market through the company's video operations unit. This is a good example of a large studio distributor working with an independently produced (or co-financed) film to fill its release schedule.

While Warners uses the same banner (or trade name) to directly distribute to the various markets available, many of the studio distributors use a specialty releasing operation for independent and acquired product. Fine Line is the independent arm of New Line, and is responsible for releases such as *Gummo*, *Hoop Dreams*, and *Shine*. Fox Searchlight is Fox's branch for indie product, offering *The Full Monty*, *One Hour Photo* and *Antwone Fisher* to theatergoers.

Through both the size of the companies and scale of productions released into the marketplace, studio distributors command the industry. Most have long histories in the business and can get product into every market and territory by the most efficient means possible. Studio distributors usually deal with every other kind of distributor on the list to both locate and release a steady stream of films.

Independent Distributors

As the name implies, independent distributors operate independently of the traditional Hollywood "system." While most of their business takes place in the home entertainment industry, independent distributors do release films theatrically — though on a much smaller scale than their studio brethren. Some of the most active independent distributors include:

- New Yorker Films
- Lions Gate

- Creative Light Worldwide
- Concorde/New Horizons
- Zeitgeist Films
- Xenon Entertainment
- York Entertainment
- Strand Releasing
- Shooting Gallery
- Phaedra Cinema
- Trimark
- Studio
- First Look
- MTI
- Anchor Bay Entertainment
- Full Moon

An independent can specialize in a particular product line — such as Full Moon does with horror and fantasy films — or distribute movies of all genres like York's offering of action, horror, and urban motion pictures. Independents can also specialize with regard to their markets and territories. Phaedra Cinema is a very small, boutique theatrical distributor that places its films in small art movie houses. Net-based *iFilm.com* distributes exclusively via the Internet.

When we began selling *Pig* to libraries in our area, 75 locations purchased approximately 100 units of the movie on VHS. Realizing this was a viable marketplace, we did some research and found that the majority of libraries across the country buy their videos and DVDs from Midwest Tapes, an independent distributor specializing in this market.

Like the large studio operations, many independents also finance and produce their own product in addition to acquiring films from independent producers. They will work all channels of the marketplace — from domestic video/DVD releases to Internet streaming to PPV deals and broadcast sales — to maximize profit potential of any product released.

Subdistributors
Though independent in nature, subdistributors are categorized separately because they do not really produce or acquire any product. An independent distributor can function as a subdistributor, moving product through their already established "pipeline." But strict subdistributors

are really only middlemen, rarely owning rights or licenses to product, just merely acting as resellers.

Subdistributors are utilized by most every studio and independent distributor in existence. Getting product to niche, foreign, and retail markets makes up the bulk of their work. Also known as sales agents, subdistributors "buy" product from many sources: large studios, studio-based distributors, small specialty independents, even producers themselves wily enough to locate and work with these companies. *Killer Nerd* was primarily distributed by the now-defunct Hollywood Home Entertainment, a direct-to-video releasing company. In some deals, Hollywood Home Entertainment used the services of a company called Mayfair Releasing to move large numbers of tapes into specific niches of the home video marketplace. Additionally, sales were made to some subdistributors directly, who then sold the movie to both rental and retail locations.

This is where the business of selling and distributing your own films and videos gets complicated. With a recognized studio or independent distributor, you will likely license or sell the rights to your project and get paid on a quarterly basis. All the details involved in contracting with subdistributors — packaging, shipping, delivery items, and pricing — are handled by company staff and are "invisible" to you. You get to move on to another project, knowing your baby is being handled competently and honestly (or so you hope). Selling directly to subdistributors means handling all of these matters yourself, while at the same time trying to produce another film and possibly maintain a full-time job and semi-normal life. Tricky to say the least. And then there's the money issue. Unlike a straight acquisition deal, subdistributors usually buy on terms or C.O.D. (which is preferable), meaning you have to front-end fund the creation, duplication, packaging, and shipping costs without being paid for possibly up to 120 days later.

Regardless, subdistributors, like other distributors in the business, have many connections to markets and buyers. This group is especially important for the foreign marketplace, where it is tough to broker a sale on your own. Each country has its own policies and practices with regard to film purchasing. Subdistributors specialize in one or several foreign marketplaces and know the "lay of the land." This kind of knowledge — long and tough for an independent producer to acquire without the benefit of prior experience — allows those just getting started in the entertainment business to actively participate in international deals.

Wholesalers

Plying their trade in the video/DVD markets only, wholesale distributors can be categorized in four distinct groups according to their products and services; adult, new product, used product, and leased product. Like other wholesalers, video distribution wholesalers are middlemen who move large quantities of product from suppliers to stores. The suppliers in this case can be studios, independent producers, video stores, or anyone who has an inventory or line of programming to sell.

Adult wholesalers obviously deal with adult-themed programming. There are a slew of regulations describing how and where this product (new and used) can be sold and displayed, and it's the wholesalers' business to manage inventory within these restrictions. Adult wholesalers will often purchase an entire previously viewed adult library from a video store (or stores) eliminating that kind of product, and resell the batch of tapes to other stores still offering adult titles for rental and purchase.

New product wholesalers are the lifeblood of the home entertainment industry. Companies such as Ingram, Baker & Taylor, and Video Products Distributors (VPD) provide video retailers and rentailers with access to the thousands of videos and DVDs available from many different sources, including all the titles released by the major Hollywood studios, independent film and video companies, and many other producers of movies, fitness, instructional, educational, and other video products. Unlike the adult-specific wholesalers, new product wholesalers do not sell to the public and operate in business-to-business situations only.

In addition to offering tapes and DVDs, many new product wholesalers provide additional services to their customers, including Web site design and maintenance, poster packages, and other sales- and rental-enhancing marketing and merchandising tools.

In the past five years, the new product wholesaling industry has experienced a shakeout. Many of the large existing new product wholesalers have become what are known as "category killers," effectively operating on all wholesaling fronts and eliminating much of the competition. Companies that at one time offered a single product line or service are now buying, selling, and leasing new and used general interest and adult product. Additionally, many of the large studios are starting to deal direct with the retailers — skipping the services of a wholesaler. This process allows for lower prices for the buyer and increased profits to the studio.

Used product wholesalers specialize in buying and selling excess and under-performing video inventory. This stock often comes from bankrupt video rental facilities. Also known as "secondary" distributors, most used product wholesalers work specific market niches like dollar chains, warehouse clubs, rackers, truck stops, and flea markets. Florida-based Distribution Video and Audio, which moves in excess of 10 million tapes per year, resells previously viewed (and now, new) VHS cassettes and DVDs to government and military installations, including public libraries, embassies, and domestic and foreign bases.

Working in the closeout arena, used wholesalers do massive liquidations from the studios, duplicators, bankruptcies, mass merchants, and other retailers for cash settlements. Videos and DVDs in these deals can be retail returns, overruns, discontinued items, and cutouts, with quantities usually in excess of 1,000 units or more. Care is taken with used product deals to protect the suppliers' and retailers' current marketing and territory deals by only distributing to channels that will not interfere with primary sales efforts. For example, if a discontinued karate video is being sold through domestic mass merchandisers like Wal-Mart and Costco, that same tape will not be marketed to dollar stores for a lower cost.

Even though the major wholesalers want to and do sell independent product, producers must often deal with a secondary buyer before reaching the main players in this category. Companies like Global Entertainment and Pro-Active Entertainment serve as independent clearinghouses — providing an avenue for small-scale projects to reach the large wholesalers. These firms provide many services to the independent (such as securing a UPC) and do much of the legwork required to get a product in shape before selling through the wholesale chain. Just as large retailers don't want to buy from hundreds of different vendors, neither do Ingram and VPD. These secondary wholesalers work with multiple independent suppliers, enabling them to sell a line of product to the primary wholesalers at once.

Leased product wholesaler Rentrak distributes videos and DVDs using a revenue-sharing model called Pay-Per-Transaction (or PPT®) that the company pioneered. Using the PPT system, more than 10,000 video outlets throughout the U.S. and Canada acquire videocassettes and DVDs from major studios and other suppliers at a much lower cost in return for sharing a portion of their rental revenue with the supplier. In this set-up, a retailer will pay from $0 to $8 upfront for the same movie or video that would have cost up to $65 from a traditional supplier. After all fees and splits, the retailer then keeps about 40% to 60% of the rental revenue. Rentrak distributes product for many of the major studios, including Disney, Paramount, Fox, Universal, MGM, and DreamWorks, as well as 30 other video suppliers that release substantially all of their rental-priced titles to Rentrak. In all, Rentrak supplies approximately 60% of the total titles released through distributors each year.

Once a novelty, Rentrak's leasing option has become very popular with small video stores because lower cassette costs allow for less risk, quicker breakeven, and greater cash flow available to stock additional copies of popular movies plus a wider variety of other titles. Other leasing companies have sprouted up, such as Florida-based ML Entertainment, which leases videos and DVDs to retailers for short 6- and 8-week terms.

Leasing programs are also attractive to the independent film producer. Rentrak offered *Killer Nerd* through their catalog and 600 units were leased to video stores. On the front-end of the deal, we only made $6 per tape (earning us a very small profit after duplication and packaging costs). During the 2-year duration of the program, however, we earned $.25 cents every time each one of those 600 tapes rented. And, at the end of the contract period, the company purchased the tape outright for another $6. At the time, Rentrak was willing to deal with single title producers to help bolster their catalog and launch their service. Though they still deal with quality independent products, the successful company is now more interested in making deals with product line suppliers.

Chapter 7 Summary Points

- Working with a distributor allows you insider access to a confusing industry that is often inaccessible to newcomers.

- Distributors exploit the marketplace through ordered selling — releasing a project to the most lucrative audiences first.

- As the number of products, markets, and audiences available has grown, so has the number of distributors servicing them.

- Studio distributors like Warners and Columbia TriStar acquire independently produced feature films to fill out their annual release schedules.

- Independent distributors often specialize with a particular product line that caters to niche markets.

- Subdistributors never really acquire product, but act instead as an intermediary, moving films and videos through already established distribution channels.

- Wholesalers move large quantities of new and used tapes and DVDs from suppliers (producers, studios, and others) to retail and rental stores.

DISTRIBUTOR FUNCTIONS

Distributors distribute films and videos. But just what does this process of getting your project to an audience entail? First understand that distributors are businessmen. They may love art, but they are not artists. They are in business to make money. Every decision they make regarding whether or not they will work with a film or video has to do with money, not cinematic excellence. You want a distributor to have this quality if you are basing some part of your future on their efforts. I've had conversations with distributors who loved *Pig*, but didn't want to acquire the film. They liked it as a movie, but knew their company couldn't successfully find a market for it. Meaning, they couldn't make money distributing the film. And if they weren't going to make money, neither was I. That's fine. I'd rather someone tell me "no" than unprofitably tie up the movie for a couple of years just for the satisfaction of saying I landed a distributor.

Obviously, distributors watch a lot of films. Well, to be more exact, they watch some of a lot of films. It's a rare day when a distributor doesn't spend some of his time either in a theatre or sitting in front of a television screening a potential acquisition. It's the only way to find product to sell. Everyone on a distribution company's staff may not view the latest product they're hawking. But I have yet to meet a distributor who wasn't intimately familiar with every film he was representing to the various markets.

ACCEPT OR REJECT CRITERIA
After watching a film or video, the distributor will choose either to acquire the project for distribution in some media (theatrical, home video, television) or reject it. These decisions are based on a variety of both objective and subjective criteria, including:

- Presence/absence of stars
- Production quality (picture and sound)
- Amount of violence

- Amount of sexual content
- Program length
- Format (film vs. video)
- Genre
- Soundtrack
- Packaging/art
- Market conditions
- Perceived appeal to current audiences

Keep in mind that distributors are human, and like the rest of us, they have bad days. Unfortunately, potentially successful films and videos can find their way to the reject pile because of the timing of their solicitation. You don't want your film screened after a distributor finds out his best salesman is quitting.

GEARING UP THE MACHINE

Once the distributor selects a film or video program, and an acceptable contract is offered to the producers, the marketing machine starts to crank. Depending on the size of the company, several activities are initiated at once. For studio and larger independent distributors, creation of a promotional campaign is the first step. Key art, posters, logos, trailers, and other media support materials used to promote the film to the public are developed by an in-house or outside agency. Publicity also begins at this point: writing, arranging, and planning press releases, interviews, video press kits, and premieres. Key entertainment contacts are targeted (film critics, entertainment magazine editors, television producers) to arrange exclusive promotional opportunities.

At the same time, business-to-business materials are crafted. These are the tools used to sell the film or video to theaters, video stores, retail outlets, Web sites, broadcast and cable stations. On the video side, this includes sell sheets, catalog pages, program incentives, store posters, shelf-talkers, and customer give-aways. Anything that can be used to create excitement about a film or video and demonstrate its potential appeal to an audience is used.

Early in the process the distributor will get the sales staff on line by promoting the project to them. They will ultimately be responsible for the selling of the movie to its various markets, so their support of the product is crucial. Screeners, fact sheets, and detailed pricing and incentive information (for both the buyer and the seller) are some of the basic materials provided salespeople.

MAJOR TASKS

The business of calling and meeting personally with potential buyers, screening the film, and finalizing sales is the bulk of the distributor's job. Theatrical sales involve screenings for exhibitors (the people who own the movie theatre chains). Stars of the film often attend these screenings to mingle with the exhibitors in an attempt to boost enthusiasm for the project. Video sales also include screening situations, but only for big distributors courting big buyers (like large rental and retail chains) on big projects. National conventions and film markets also provide sales forums that allows for distributors to scout out sales. Film festivals in particular are an increasingly important showplace for new product. Distributors frequent these around-the-calendar and around-the-globe affairs courting organizers, soliciting buyers, promoting projects, arranging sales, and looking for their next acquisition.

With a small independent distributor, mailings and follow-up phone calls (and some personal selling) get the job accomplished. As it's too costly to send full-length screeners to every potential buyer, new product is sold to video stores on the basis of its art, reviews, publicity, and price.

Advertising campaigns, both trade and consumer, are planned in this time. Decisions on type (print, radio, broadcast), size and length, look (color, black-and-white, content) and schedules are influenced by the forecasted profit potential of the project. This forecast is created from distributor-conducted research on potential audience size, test screenings, screen (or shelf space) availability, competitive products, economic situations, and — sometimes — gut feelings.

ANCILLARY SUPPORT

In addition to the basic sales and marketing work involved in creating theatrical, video, broadcast, and Internet sales and placement deals, distributors provide an array of ancillary support work that accompanies the effort:

- Shipping – moving film prints or video copies from studios, warehouse, and duplication facilities to their various destinations (includes boxing, packaging, labeling, postage, and tracking)
- Collections and Payments – having the clout to make sure monies due are received and paid out in a timely fashion
- Monitoring – developing and maintaining reports on sales, and tracking screenings and placements of the project in all markets and media
- Subtitling/Dubbing – creating subtitle or new dialogue tracks for film sales to foreign markets
- Editing – making necessary cuts to both video and audio content for ratings, censorship, and program length constraints
- Rating – getting the film or video approved by the Motion Picture Association of America, a sometimes lengthy and always costly process

Again, the execution of many of these tasks is dependent upon the distributor and contract. Working with a studio distributor is a turnkey operation. You give them your project, they give you an audience. The independent producer dealing with a small distributor is another story. Many of the tasks outlined above will fall to the film or videomaker. And for those taking the self-distribution route, you've just had your first glimpse at what lies ahead.

Chapter 8 Summary Points

- Most every decision a distributor makes regarding a film or video is based on that project's potential profitability.

- Distributors choose whether to acquire or reject a project based on a number of items, including star power, production value, genre, format, sex, and violence content.

- Creation of a promotional campaign is often the first and most important task performed by a distributor.

- Personal selling, film screenings, advertising, shipping, collections, dubbing, packaging, and sometimes even editing can all be distributor responsibilities.

HOW DO YOU LAND ONE?

Imagine for a moment that a distributor has heard of your project. He calls, offering to represent the film or video throughout the world and all its markets and media. A deal is struck and his efforts ultimately secure sales and fame for both the project and you.

That's what's called a best-case scenario. Others would refer to it as a dream. Is it possible? Yes. Is it probable? Well, it's happened. But honestly, the odds of a distributor approaching you, without any sort of formal solicitation on your part, are not the kind you want to play in Vegas.

When it comes time to finding a distributor for your project, the responsibility, and most, if not all, of the work involved in the process will fall to you. It's not going to be easy. It's not going to be quick. It's not going to be painless. It is, however, going to be worth your trouble.

Luckily, securing the services of a distributor isn't always completely dependent on the "who-you-know" philosophy that seemingly permeates every deal in the entertainment industry. Undeniably, who you know helps. Connections open doors. Truth be told, personal favors have made careers. But in the acquisitions phase (especially for those with finished product), you do have an advantage because profit-driven distributors always want to befriend the "next big thing." And that's exactly what you'll have the possibility of becoming in the minds of everyone you approach.

Who could've predicted the popularity of *The Blair Witch Project* or the Tae Bo phenomenon? Breakout independent success stories such as these fuel acquisitions paranoia in the film and video industry because it is every distributors dream to be attached to one. And for all anyone knows, you are the next "one."

Marketability of the project is another reason the who-you-know philosophy doesn't always affect the acquisitions process. Granted, if a distributor needs a drama, and it's between your non-represented submission and another similar-quality film that comes with an industry-insider attachment (such as the buyer's friend or associate), you'll lose. But just because your second cousin Maryann is the secretary to the director of acquisitions at Paramount, it doesn't mean he's going to put his job on the line promoting your unmarketable film. It doesn't matter who you know (or quasi-sorta know) if your film or video doesn't hold promise for audience acceptance, or worse, sucks. You will get some polite and possibly very helpful advice, but no deals.

QUALIFYING DISTRIBUTORS

Like all other marketing tactics detailed in this text, the search for a distributor should start when you begin planning your movie. Ideally, at the earliest stages, you want to create a project that meets distribution needs and expectations. That way, you avoid the danger of investing too much time or money developing a potentially unmarketable product. The task of qualifying distributors according to their current and future market needs, and *then* designing a film or video to match those needs, often conflicts with the "artist" side of independent film and videomakers. Realize, however, you may lose all of your money, and more importantly, your "shot" by ignoring this kind of front-end, marketing-focused work.

Suppose you've written (what you believe to be) a great horror movie script along the lines of *I Know What You Did Last Summer*. Using $100,000 (your life savings), you jump into production, knowing the film will be a sure-fire hit. When you start shopping the finished project to distributors, you're told that "pretty-people-slasher flicks" have run their course, and the best offer you can find is for half the production cost. This is when being an artist isn't so much fun. Unless you enjoy the starving part.

SELLING OUT?

I'm not saying to ignore the artistic expression involved in making a movie or video production. Just don't let it get in the way of commerce. If your creative passions and interests run counter to distributors' (as well as the public's) wants and needs, then you should consider film and videomaking as an art form, not a career. There's nothing wrong with that. Trying to sell something to an unresponsive audience is depressing. I remember shucking candy bars in front of a grocery store in seventh grade. It wasn't fun. And if you cannot view your future film or video as a hook-laden marketable product, and realize that this part of the process is all about making money, then your search for a distributor will be just as joyless.s

If you've already completed a project, qualifying distributors is still important. In addition to saving you time and money when you begin the solicitation process, this task provides the opportunity to shape your sales and promotional efforts around already present elements in your project that distributors find attractive.

Again, this may sound like you're selling out. That may be true. Remember though that distribution is a business. There is a product to sell. Whatever you can do to make it easier for a distributor to say "yes" needs to be done. You need a distribution deal to launch your career, to get established, to be a reliable investment when you look for cash for your second film or video. Do you want to find yourself in the position of owning a finished production that nobody wants to touch? The only way to avoid such a fate is to find out what distributors want and give it to them. It really is that simple.

The process of qualifying distributors for independent film and video projects involves four basic steps:

1. Collecting distributor information
2. Assessing potential distributors
3. Refining your database
4. Designing an approach plan

This progression of collecting and examining the information gathered in the process and applying it to your marketing efforts allows for an efficient and effective use of both your time and money — enabling you to approach a more targeted audience with a product they are more likely to buy.

COLLECTION OF DISTRIBUTOR INFORMATION

Before you begin anything, you need a list of companies that distribute entertainment product. Where do you start? One place is the Appendix of this book (see page 342). You'll find a list of independent-friendly operators that range from small video distributors to large studio players. These companies have a history of working with independent projects in a variety of genres, program lengths, formats, and budgets. This list is not all-inclusive and may contain inaccurate data by the time this is printed. It is a great starting point, but there's more out there.

Next, hit the Net. An indispensable tool for research of any kind, the Internet is extremely helpful when collecting distribution company data. Using Google, Yahoo, or any of the other popular search engines, try the following keyword searches: motion picture distribution, video distribution, film distribution, independent film distribution, movie distributors, motion picture distribution companies... you get the idea. Some of the more useful sites are again listed in the Appendix of this book. As you surf distributor sites, keep a notebook handy, or open a blank Excel file, so you can record Web addresses of those locations that appear to contain useful data.

Now it's time to visit a couple of video rental and retail stores. Check the back of video and DVD packaging for distribution information. With that notebook in hand, copy the company name, city, and phone number that is provided on most products. Do yourself a favor and talk to a manager first — alerting him to your actions. It's not illegal to do this, but very suspicious. How many times have you seen someone in the video store taking notes?

If you are planning an instructional or other non-feature project, check out retail stores and outlets that sell similar items. Hardware merchandisers carry a full line of do-it-yourself home repair tapes; sporting goods stores sell fitness tapes; book and gift shops offer spirituality, wellness, and self-improvement programming.

Check filmmaking associations and groups such as the Association of Independent Video and Filmmakers (AIVF), Independent Feature Project (IFP), and others. These organizations exist to offer support and resources to independent artists. Most have research libraries (both physical and online), monthly publications, and international member-ships for networking. Talking to both established and aspiring film and videomakers is a great way to gather information on independent-friendly distributors. Also search out the American Film Marketing Association (*www.afma.com* or 310-446-1000). A trade group for studios and dis-tributors, they offer very comprehensive and up-to-date information on hundreds of firms that specialize solely in the business of film, video, broadcast, and cable distribution.

Peruse libraries, bookstores, and newsstands. *The Hollywood Reporter, Variety, Moviemaker, Independent Filmmaker,* and *The Independent* are all newsstand publications that offer distributor list-ings throughout the year. With the fore-mentioned *Reporter* and *Variety,* focus on those issues printed a week before and the week of a sales market (American Film Market, Cannes), as these magazines become the film buyers' "bibles" and will be heavy with distributor advertising.

Trade publications not found on the newsstand also offer a great source of distributor information. These are the magazines that people "in the business" read to get news on the industry. The video industry is covered by two periodicals — contact each for a sample issue and subscription:

Video Store Magazine – (800) 854-3112
Video Business – (800) 786-0609

Both offer special distributor listing issues, tons of distributor ads, and education on the business in general.

Obviously a vast number of books have been printed on the subject of independent film and videomaking. Check out their appendices for lists similar to the one found in this text. Be aware of the time-sensitive nature of company data. Film and video distribution is a fickle and tough business — companies come and go with the breeze — so list-ings can be inaccurate (you'll quickly find that out in the next step!). In addition to traditional booksellers, look to college bookstores for spe-cialty texts on the subject.

If you have some money to spare, mailing-list brokers can collect contact data on distributors and deliver it in computer-friendly format.
Hundreds of such companies exist and most pull their information from
the same sources. One of the benefits of using such a service is that
information will be guaranteed "fresh," meaning that if a company is
out of business or the information provided is not current, corrections will
be made free of charge. Prices are based on the type of names (business or consumer), number of names (1,000 names vs. 7,000 names),
type of selections (fax numbers only, credit-rating code) and output media
(e.g., download, diskette, mailing labels, etc.) requested. As an example,
a list of 259 motion picture distributors is available from *infoUSA.com* for
$207.20. This will get you detailed company name, personal contact(s),
address, phone, fax, and e-mail information on a CDROM disc.

Finally, don't forget specialty directories and phone books. It sounds
awfully obvious, but the Yellow Pages directory was the last place I
looked when I started my first search for a distributor (and three firms
were listed within my metropolitan area!).

Assessing Potential Distributors

If you are a somewhat comprehensive researcher, then chances are
good you've got a huge batch of notebook paper, ripped out magazine
pages, photocopies, and maybe even a napkin or two containing distribution company information. Congratulations. The first step in your
hunt is done. Now it's time to really get to work.

You need to determine which distributors among those located in your
initial search seem most suited to working with your film or video. Is
Paramount really an option? Or do they just deal with multi-million-dollar
"A" movies? What about Artisan? They have a history of independent
releases, right? So what would be keeping them from working with your
film (even though the company was deluged with submissions after
their release of *Blair Witch*)? And could that small theatrical distributor
— the one you discovered through an e-mail conversation with another
independent filmmaker — really get your movie on the big screen?
Surveying potential distributors is the only way to get answers to these
questions and collect the kind of data that will lead you to make

informed and intelligent choices when you begin calling on companies in an effort to sell your film or video.

Gathering Data

Though personal contact is the best way to gather these facts, a good base of information on probably every distributor unearthed in the last step can be garnered from some more intensive Web-based research. Most, if not all film and video distribution companies have a Web presence describing the firm's history, key officers, types of product, current releases, and even submission guidelines for independent producers. In addition to QuickTime video clips, an exhaustive company synopsis, current news, and an industry link page, Creative Light Worldwide's site (*www.creativelightworldwide.com*) offers a goldmine of information in the FAQ area, with answers provided to questions like:

- What genre of film should I make?
- What sells the best?
- What is the sales process and how soon can I see money?
- Can my film get booked in theaters?

While surfing the various distribution company sites, create an Excel or other database file allowing you to organize information by name, product offerings, contact person, size, and other pertinent details as you find them. You want the data in an easy-to-access format for verification purposes, which is the next step.

Finding the correct contact person within a distribution organization, determining their needs, verifying phone, fax, and mailing information gleaned from the Web, even discovering if the company is still doing business... all of this requires more than just a couple clicks of the mouse. You're going to have to make some calls. Actually, you're going to have to make a lot of calls.

Making the Contact Call

There is no way to know what kind of product a distributor is acquiring and selling without talking to them. Distribution is a quirky business, and speaking directly to someone in charge offers several benefits. For example, a company that specializes in children's films (and initially appears to have no interest in your exercise video) may also, unbeknownst to you from surface research, market a line of instructional tapes under a different brand name. Beyond opening up new market opportunities, initial telephone contact saves you money. Verifying which firms fit the "potential buyer profile" established for your film or video project is usually only accomplished through personal interaction. What at first may seem a perfect match (a distributor's Web site promotes a line of independent horror films) can quickly not seem so "right" (the horror films advertised all include soft-core "action") following an enlightening telephone conversation. E-mail is also an option, though it is always better to speak with someone personally. It's a lot tougher to dismiss a request from a live human being than an electronic message.

Making the calls is like doing an interview. You want to gather the most amount of information by the least offensive means possible. Distributors are busy folks. Don't waste their (or your) time.

Obviously, the first step is to phone potential distributors. You'll want to ask for someone in acquisitions. If you were able to glean this name from your initial research, great. If not, just ask for new product acquisitions. Be professional and extremely polite to whoever answers the phone. She will be your only contact at this point, and believe me, this "gatekeeper" holds a lot of power. She'll ask for your name and the company you represent. Simple enough.

More often than not, you'll be routed to a general acquisitions office, where a second-level assistant will answer the call. This individual may have you quickly explain your project before handing off the call to someone in charge. If your film or video sounds appropriate (guess what, *you* are getting surveyed now!), you will get to speak with an acquisitions agent or director. Perfect. This person can offer you the most accurate information about their business and what kind of product they need. Be ready to offer a short description of your planned or completed project, before you start asking any questions.

A list of the kind of information you want to have on hand during a distributor contact call:

1. Concept capsule
2. Audience profile
3. Genre description
4. Cast highlights
5. Length of project
6. Reviews/awards/festivals
7. Marketing support
8. Budget (if pushed)...

While you are not selling your project in this call — just gathering facts and gauging interest — you do want to play up the merits of your film or video. Describe it in realistic, yet favorable terms, comparing it to similar, successful projects in its specific genre or field. It may become immediately apparent that your project is not something the distributor would handle — that's fine. That simple phone call just saved you some money and the distributor some time. These research calls also help to avoid the embarrassing situation of soliciting a completely inappropriate company with your product, situations that leave you looking unprepared and unprofessional.

A high point is that buyers want to talk with producers. They are always looking for new product and people like you are their only source. Most acquisitions folks are really nice, willing to answer your questions at length, though some of this responsiveness may be fueled by the fact they are afraid of passing up a potentially profitable product. Another plus: The phone call helps to establish a personal connection, one that can be used to elevate your solicitation above others that may have been submitted "blind."

Approach these **surveying calls** in a systematic manner, determining a certain amount of time or a specific number of calls you'll make each day until you've spoken to everyone on your list. Before picking up the phone, prepare a script of questions for potential distributors that should include:

- Do they accept unsolicited or unrepresented submissions? Will they even look at a project if it didn't come to them through official channels?
- What type of product does the distributor handle? Are they selling instructional tapes or music videos, urban dramas or sci-fi animation?
- What genres do the best business for the distributor? For example, a company may sell everything but have a great track record with reality-based videos.
- Where do they market their films and videos? Retail and rental stores? Mass merchandisers? Catalogs? Foreign buyers? Television, cable, or the Internet? Do they sell through sub-distributors?
- What kind of marketing do they use to support their titles? In-store posters? Trade or consumer magazine advertising? Radio, TV, or Internet commercials? Do they create their own or work with outside publicity firms?
- What kind of budgets are they looking for? A distributor will often classify projects by dollar amount. They are not so much interested in if you have a western or a romance, but whether or not you spent at least $1 million making the film.
- What are their submission processes and policies? At what point do they like to assess a project — before, during, or after it is complete? Does the film need star power? Can it play in festivals? Is it color or black-and-white? What formats are acceptable? Are there length and soundtrack requirements?
- What product does the company currently offer? Are their movies and videos popular with their target audiences?
- Do they have any available marketing materials? It's interesting to investigate a company's ads, sell sheets, catalogs, packaging, press releases, and other promotional items to gain insight on their marketing techniques.

If time permits, and the conversation is friendly, you might also inquire about the kinds of deals they have offered other producers whose product they are now selling. Much of this answer depends on your specific project, its stage of production, financing arrangements, and a hundred other details. But the bottom line to all these conversations is money, so it's nice to get an idea of what you might expect from any potential deal.

Several minutes into any conversation, it'll become obvious if a company represents a likely distribution partner for your film or video. If so, let the acquisitions representative know that you'll be soliciting them soon, and inquire as to whom the package should be sent. Follow up every call with a thank-you letter. Yes, send these to *everyone* — even the companies that are obviously not interested or not a good match for your project. The easiest method is to use a thank-you form letter that can be personalized in one or two places. Mail it that same day. You want top-of-mind awareness for your project, and the more contact you have with distributors, the better your chances of this occurring.

REFINING THE DATABASE

Two things probably became obvious in the last step: 1) There are a lot of distribution companies out there and, 2) There are a lot of distribution companies that *could* be interested in your film or video. While this may be true, you don't want to waste your time and resources sending your project to everyone. This "shotgun" approach will only cost you a lot of money and result in a lot of rejection. Traditional distributors are not always interested in independent productions, no matter what they tell you. You may have created a product that is difficult to market or concerns too specialized a topic, yet distributors still want to view it, "just in case." Remember, you might be that next big thing...

And don't forget: Your goal is to identify distributors that carry films or videos comparable to your own. This practice will provide you the best chance of success. Don't believe that if a distributor already represents a product similar to your own that they won't be interested. Quite the opposite is true. If a distributor has five horror films in their catalog, that means they have created successful relationships with buyers interested in horror product. They will continually need new horror films to offer these buyers. So if you're promoting a horror feature — they're your guys!

What you need to do is refine the database of information you have just collected in a way that allows you to identify those companies most likely to say yes to your solicitation. Examine your list of distributors, paying special attention to those companies that:

- Indicated interest in your project during a personal conversation
- Currently work with films or videos in the same genre/subject as your project

■ Currently work with films or videos with production values and budgets similar to your project

Companies that meet these criteria will be the first you solicit for a distribution deal, as they are most experienced in successfully distributing and selling product similar to your own. They are comfortable with what you have to offer — basically they know what to expect.

Guess what? After picking out the cherries, there are still a lot of possible distributors left on your list. Sift through your database a second time, looking for distributors that offer at least two of the above characteristics. For example, an acquisitions executive at York Entertainment may be very interested in your urban-action flick, as the company has a release schedule full of such product. Their movies, however, may boast higher budgets and minor stars. Regardless, they represent a solid prospect, and will be contacted if you find no success with your first round of solicitations.

DESIGNING AN APPROACH PLAN

You want to be smart about this "thinning-out" process, eliminating the obvious rejections before they happen. If you do your homework right, there should be no surprises when a distribution firm receives your screener and presentation package. Some points to consider:

Be Realistic About Your Project

When we made *Pig*, there was a belief that it could capture a *Blair Witch*-sized audience. Our rationale: Both movies were independent reality-based features. But that wasn't being honest with ourselves. *Blair Witch* is a good ole' fashioned scary movie whereas *Pig* is a lot tougher to categorize (therefore a lot tougher to market). And don't waste time waiting for Paramount to respond to your shot-on-video, Hot Wheel® toy car collecting documentary. Big studio and large independent distributors need to release films that can gross millions of dollars for them theatrically and in video/DVD markets.

Don't Exaggerate the Merits of Your Project

Be honest when discussing your film or video with an acquisitions executive. Sure, sales savvy is important. You're trying to sell something so you have to hype it. Just don't lie. When setting up distribution for *Killer Nerd*, I was often asked about "star value" of the film. We had an MTV personality, so I wasn't lying when I said we had some celebrity. But I clearly explained who the actor was and what his role on the music channel had been to that point. Distributors won't be pleased if you drop recognizable names and big-budget figures only to later see a shot-on-VHS backyard movie starring your immediate (and not famous) family.

Approach the Approachable

People will tell you to dream big. And you should. But when it comes to finding a distributor, sometimes a smaller company specializing in a niche market can offer you and your project the best opportunities. I know it would be pretty cool to have the Paramount logo come up on the screen before your movie begins, but you may find during the qualification process that a big studio distributor just isn't a good match — regardless of how you massage the information. A small company may be better able to target and reach the audience and markets appropriate for your film or video production. There's something to be said for being the "right sized fish in the right sized pond."

Make a Good First Impression

In plain English, don't be rude. A business-like approach to any communications, be it personal call, voice-mail, e-mail, fax, or letter, is best. More often than not, you won't be meeting these folks in person, so the only things they have to go on are your communication talents. If you don't have the skills to be a polished presenter, find someone who does and make them your "salesperson." Always err on the polite side. This goes double when speaking to secretaries, assistants, and phone operators. Like any business, distribution firms are tight-knit communities, and if you are a jerk or behave in a non-professional manner to someone answering the phone, believe me, everyone working there will know about it — including the acquisitions people. Remember, these people may be responsible for the future of your career. It is much easier for them to ignore or reject your solicitation than it is to show interest. Please, don't make a negative response any easier.

Chapter 9 Summary Points

- Most, if not all, of the work involved in finding a distributor is the independent producer's responsibility.

- Fortunately, gaining distribution representation isn't entirely dependent on "who you know."

- In a perfect world, you would qualify distributors as to their current and future market needs and design a film or video to match those needs.

- Gather as much information on as many distributors as possible before you begin shopping your film or video around.

- Look to the Internet, video stores, trade magazines, industry organizations, and even the phone book for a wealth of data on film and video distributors.

- Personal interaction with distributors is the best way to confirm any potential interest they may have in your project.

- When placing the "first-contact" call, quickly describe your project and ask questions that will lead you to discover if the company is a possible match for what you are offering.

- You don't want to solicit every company that seems interested in your project.

- Revise and refine your list of potential buyers to make your solicitation efforts more effective.

- Be realistic and go after those distributors best suited for the characteristics of your project.

WHO AND WHAT HELPS
IN THE PROCESS?

Just as a distributor adds clout when it comes time to selling your film or video to the various markets, so too can an industry representative — or venue — add clout when you are hunting for a distributor. Having your project presented by an industry insider or viewed in a favorable environment breaks many of the barriers facing an independent's entry into the market.

Some may believe that a distributor has a charmed life. Not so. They watch upwards of 40 movies and videos a week, and the better part of those submissions is trash. That's in addition to all the travel and promotional work a distributor does for products they already represent. All they want to do is find the next gem, that needle in a haystack. And if someone can point the way, all the better. Getting your film to the top of that heap — and surviving — is the goal. The following people and places can facilitate that process:

PRODUCER REPRESENTATIVES

These folks deal with distributors on a daily basis; their connections are obviously beneficial to your project. Let's say you are hunting for a video to watch on Friday night. Who would you go to for a recommendation — a stranger or a business associate who knows your tastes and interests? The same rationale is at work here. It's human nature to respond favorably to someone you know. To trust their judgment. To believe they want to help you.

A producer representative's job may seem sketchy, with all the party-going, travel, and non-tangible work effort, but it's the nature of the business. Their whole gig is to connect people and product and markets. If your film or video is part of that networking loop, then your chances of getting noticed by a distributor rise dramatically.

You would approach a distributor's rep in much the same way you would approach a distributor. Collect information on working and available producer reps, survey these individuals as to their wants and needs, weed out those that offer the most promise, and contact them with your film or video and support materials.

Producer reps get involved with projects they believe will find an audience, popularity, and above all else, solid sales. Since most producer reps are self-employed specialists, they stake their reputation on each film or video they represent. A clunker or two is acceptable over the course of several years' worth of movies. But if these guys continually promote bad films and unmarketable videos to their distribution contacts, they'll soon find their business deals going south. In many cases, a producer reps' only source of income is the percentage she's making off any sales deal she arranges. The better the film or video, the better the deal, the better the percentage. Some reps also work on a fee-basis. You pay them to find you a distribution deal, festival slot, or other exit out of independent film and videomaking obscurity.

The downside to working with a producer rep is that percentage or initial outlay of cash. But if you can afford the costs, it's money well spent. It'll be much harder, if not impossible, for you to replicate the role of a producer rep. They work to create and stimulate interest and excitement about your project, maximize that interest, with an end goal of creating competition among distributors and other buyers. That's quite a challenge for an independent filmmaker with one 75-minute, 16mm movie under his belt.

How do you know a good producer rep from a bad one?

- Track record – What kinds of films and videos has the rep worked with? How many of these projects successfully found sales and audiences? How long has the rep been in business? Is the rep also an attorney? (This is important when it comes time to negotiating deals and signing contracts.)
- Charges – Has the rep worked for a deferred percentage on films she's liked? If so, then it's probably a safe bet that the individual is a fan of movies as much as she is a fan of making money. Do up front fees and back-end percentages seem appropriate to your project?

- Opinion of others – How is the reps' rep in the industry? Does she deal with established buyers and producers? While it's preferable to work with an industry "approved" individual, the only opinions that should really count are those of other film-makers who have used the services of the rep in question. Speak to others who have utilized her services. Ask about the value of the service. A list of independent-friendly reps can be found in the Appendix (see page 354).

AGENT

Like a producer representative, one of an agent's main functions is to connect the film/videomaker with appropriate distribution contacts. Agents use many of the same tactics as producer reps in securing sales for a project, with billing occurring on a percentage or flat-fee basis. Again, you'll want to survey the hundreds of entertainment agencies available, choosing to contact only those that both accept unsolicited submissions and seem right for your project.

FILM FESTIVALS

An excellent way to get your film seen and reviewed, a film festival is also a great venue to spark the interest of a distributor. Festivals are notorious for becoming "ground zero" for a film's buzz. The right movie in front of the right audience and bang — acquisitions representatives are all over the filmmaker with distribution deals. Getting into the right festival is part of the strategy, of course. Showing your film at a small-town gala that's screening stuff barely a notch up from home videos is not going to get you any offers. But a slot at Sundance, Toronto, Slamdance, or Berlin, will. One catch: You'll need representation to better your chances. Enter the producer rep (or agent) once again. If a producer rep can personally present your film or video to a programmer with his recommendation (read sales pitch), it can make the difference between an acceptance or rejection letter. A film festival creates a positive and enthusiastic film-friendly environment that cannot be duplicated in the distributor's VCR-outfitted office screening room.

FILM MARKETS

Created for the sole purpose of selling films, videos, and other pro-gramming, film markets offer a quick link to distributors and buyers.

Like a flea market for entertainment product, events such as the American Film Market in Santa Monica, California, are populated with over 300 motion picture companies and 7,000 film executives. Acquisitions and development executives, producers, distributors, agents, attorneys, buyers, and film financiers convene to pursue the business of film. Hundreds of films are financed, packaged, licensed, and given the "greenlight," culminating in over half a billion dollars in business for both completed films and those in preproduction.

Obviously a place of action, a film market can be quite intimidating to the novice indie producer, though many attend the Independent Feature Film Market in New York as a way to drum up distributor interest in their projects. At this particular event, filmmakers can set up a sales table, rent private suites, screen their films, promote scripts, or just mingle with the throngs of people attending. It all comes with a price, which, depending on the event and location, can get pretty steep.

Producer representatives spend a good deal of their working hours at film markets around the world. If you've contracted with one, they'll use the exposure of these events to aggressively sell your movie or video. They tend to promote packages of films to offset the high cost of the affair. This tactic not only allows for greater sales volume, but also provides the rep the opportunity to offer specific product to many classes of buyers.

More detailed information on attending film markets is discussed in Chapter 17. A list of events is provided in the Appendix (see page 351).

Chapter 10 Summary Points

- A good producer rep will get involved with a project only if he believes it will find an audience, popularity, and sales.

- Look to the track record, fees, and reputation of a producer rep before signing on with one.

- Agents often function like a producer rep, aiding in the "connections" process that leads to distribution and sales.

- Film festivals and markets are good opportunities to show off your project and attract the services of a producer rep, agent, or distributor.

SOLICITING DISTRIBUTORS

Undoubtedly, your distributor search turned up many classes of buyers — from bigwig studios, direct-to-video companies, and cable outlets to theatrical releasing specialists, television broadcasters, retail jobbers, and more.

Though your goal at this point is to attract a distributor, it may be tempting, especially after considering the list of potential customers you've just collected, to solicit other buyers in the entertainment product food chain as well. Three words of advice: Don't, just yet.

FIRST ON THE HIT LIST

Plain and simple — *always contact the studio or large independent distributors first.* Unless, of course, you want to quickly eliminate any large-scale sales potential that may exist for your project. A distributor is known as such because it does one thing successfully — distributes films and videos. With an infrastructure of contacts and service providers in place, a distribution company can easily promote, take orders, duplicate, distribute, and collect monies for the sale of your project from a variety of marketplaces. Selling your film or video directly to video chains and even individual stores decreases a traditional distributor's ability of marketing your film or video into as many markets as possible, which decreases the distributor's earning potential, ultimately decreasing your chances for a deal. For example, say you've produced a nature documentary. Its high-end production values make it a natural for cable giant The Discovery Channel. That company has no problem brokering a deal directly with you, so you make the sale for quick money. After repeating this process with a couple of other broadcasters and even a video chain, you then decide to go hunting for a distributor. Do you really think anyone is going to touch the project after it has been offered on the market?

I know what you are thinking — that direct sale to Discovery cut out the distributor's percentage on the sale. But will that percentage be more than all of the sales you missed out on from not having a legitimate and well-connected distributor rep the project? I doubt it. While with the right project it's possible to make several large or obvious immediate sales, there are so many markets that you, as a producer, have no access to — buyers and media that are second-hand to a distributor, but you may not even know exist. That's the value of working with a distributor: their reach.

And you might also consider costs involved with doing your own selling. Staffs of salespeople, office space, phone lines, marketing materials, advertising budgets, financing term sales, packaging, and shipping capabilities... the list becomes endless (as you may soon learn if you start functioning as a self-distributor).

Personal selling of a film or video is tough in a market dominated by a limited number of strong distributors. Retailers are constantly stream-lining their processes, and that usually doesn't include buying individual films from every independent producer on the planet. Try to imagine if every video rental store only carried one movie each. That wouldn't make for a pleasant shopping experience, and video/DVD and pro-gramming buyers feel the same way. They want to choose from a variety of offerings that will be delivered to their specifications in a pay-ment plan that suits their individual business needs. Most independent producers shopping a single feature film can't offer that kind of service. If you're selling your film theatrically, the ballgame gets all the harder. Booking theatres on your own is a nearly impossible task these days, especially with a low-budget project. You may be able to get into a few art houses, but all the main theatres and multiplexes are booked months in advance with major releases. You'll be playing ball against the big boys — Warners, Paramount, Universal — without the help of an experienced distributor to guide you. Some have succeeded in this path, but not many.

THE QUESTION OF "WHEN?"
So now that you're convinced it's best to approach mainline distribu-tors first, the question of timing arises. At what point do you solicit a

distributor with your project? You really only have three options:

- During the project's development stage
- During production
- When the project is "complete"

Development Selling

Trying to sell ideas is a tough business regardless of the industry. Trying to sell ideas in the entertainment industry is even tougher. The reason: Everyone, from the lighting assistant to the lead actor to the producer and distribution company accountant, is selling ideas in the form of a script. Scripts are not hard to write. People who work in the industry churn them out in the thousands because it's well known that a good script can turn someone into a "player" overnight. But there's a problem with this scenario: finding that good script.

People who buy and develop ideas, including many distributors, pour through a lot of bad scripts — and the ones they do read usually come from friends and associates. Independently solicited scripts are often ignored, sent back unopened to the author (due to copyright infringement fears), or so far down the distributor's "to-do list" that it could be months before you hear a response. If time is not a factor, schlepping your script around town may find you a distribution deal. Some development companies and their executives will read unsolicited scripts direct from the budding filmmaker. Be aware, however, that in most cases, you'll need an agent or a producer rep to get your script in the right hands.

Probably a better plan: Don't send a script, but just throw your idea out there to gauge interest and marketability when surveying distributors. Remember the concept capsule? Well, if you are searching for development funds for a planned movie or video, that abbreviated description of your project will come in real handy right about now. Known in the industry as a **pitch**, selling your idea based on a couple of sentences and a five-minute conversation is actually an art form in itself. There are many books and resources on the practice of pitching, with one Web site (*www.moviepitch.com*) stating on its homepage that, "You don't need to be a good writer. You don't need an agent. You just need a great idea." Designed as an "idea-portal," the site gives non-connected independent producers the ability to try out their film and video ideas on entertainment players, in a low-cost, low-commitment manner.

Every distributor you speak to, however, is not a candidate for your pitch. Unless they specifically state development funding is a part of their service, don't waste your breath and set yourself up for unneeded rejection. You can always add the question "Do you provide development funding?" to your distributor query, and if they answer positively, have your pitch ready to go.

Production Solicitations

Completion funding is one of the main reasons independent producers contact distributors and other potential buyers during the production or postproduction phase of a project. Films and videos, even the most fiercely independent ones, are expensive to make. What seems like a sufficient budget on the first day of shooting somehow becomes wholly inadequate as the project grinds through the shooting process or nears its technology-dependent and very costly postproduction stages. To offset this imbalance of cash, producers often solicit distributors, hoping they'll become the project's "white knight." The theory is that the distribution company will be so impressed with the half-finished project, that it'll front whatever monies are needed for completion in exchange for the exclusive distribution rights of the property.

There is some danger inherent in this technique in that distributors are not filmmakers, and thus, cannot "see" the big picture. In other words, if you show an acquisitions executive your non-synched feature that's also missing optical effects and a music track, chances are he's not going to be able to "fill in the blanks" — no matter how rosy a picture you paint for him. Creating a negative image of your project at this early stage may influence a decision down the line when the film or video is in finished form.

It can be helpful to passively contact distributors during (and even before) production of your project. Mailing a news release, postcard, newsletter, e-mail, or short personal note detailing the status of your film or video is an effective technique of keeping your venture top-of-mind while at the same time maintaining contact with the individuals and companies you hope to solicit sometime in the future.

With this technique you are not actively selling your project or looking for development or completion funds. You are merely spreading the word about the film or video. In the pre-solicitation marketing of *Pig*, we used a simple two-sided newsletter that, in addition to several short articles, featured the film's key art and tagline, a short description of the plot, photos, and the official movie Web site address. This black-and-white promo piece was mailed to approximately 150 individuals, all of whom represented a distribution, festival, or other sales opportunity.

Newsletters were one of the many ways Pig *was promoted to maintain continual awareness for the project in the mind of its many audiences.*

® 2000 T.Michael Conway/Mark Steven Bosko Productions. Inc. 330-258-1098

CRITICS PRAISE PIG

THE LATEST NEWS ON THE INDEPENDENT FEATURE FILM PIG - SPRING 2000

In a recent review of the film PiG, WHIZ NBC radio's Thomas Brown declared the independent production "rich in style and content and filled with wonderful performances." Brown's complete "Eye On the Screen" review follows...

'God has a judgment day coming.' These are the ominous words that appear on a billboard located on a lonely highway somewhere in northern Ohio. Behind the billboard, sitting patiently in his car, is a police sergeant. The gentlemen looks to be in his early forties, has long reddish hair and is of a rugged stocky build. With little fanfare or emotion he tells the interviewer his criteria for gauging speed violations. At least we assume he's talking to an interviewer. We believe that we are simply observing the production of a documentary about this law officer, or perhaps it could be an episode of "Cops," but we're not sure. There could actually be something even more unsettling happening here. Perhaps we're witnessing a mentally deteriorating lawman...perhaps he's been talking to himself....and perhaps he's very close to permanent residence in a padded cell. As "Pig" makes steady progression we're not sure about much of anything... except that this film has methodically pulled us in...and we just can't shake the feeling that something positively dreadful is just around the corner.

Watching this low budget independent film I thought of other films, namely "Taxi Driver" and "The Last Picture Show." "Taxi Driver" comes to mind because it gives us an unbalanced character with virtually no definition. We can only watch in helpless gut felt suspense as the tedium of

being a New York cabby is slowly but surely pushing Travis Bickel over the edge. The character in "Pig," a small town lawman, has been experiencing a different but still extremely potent tedium as well. This is a tedium that comes from the day-to-day activities as a policeman in a very rural community... activities that are rarely...even remotely as exciting as what we see on the screen. While we're never positively sure that this incredibly thankless and stale occupation is affecting his mental state, it's certainly easy for the viewer to understand if it did.

"The Last Picture Show" came to mind because of its similarly rural location and an eerie quiet, broken up by the haunting sound of a moaning wind... a wind that seems to suggest the coming of permanent darkness... an apocalyptic type event...perhaps the apocalypse. This cop may indeed know something that we don't and he's simply biding his time. We just don't know...in fact we never know.

When it's all over, we're left uneasy. What did we just see? Who is this guy? Is he a good cop or a bad cop? Does he somehow deserve the demeaning tag of being a 'pig?' Would we feel comfortable being pulled over by this officer? Was this a real character? Is he an actor? Was he being followed around by a film crew, and if he was... why? Did something happen? Was he derelict in his duty? Did he do something outrageous in the line of duty? Did he harm someone? Did he kill someone? Is he even alive? I can't think of a movie that has ever left me reaching for more answers...deeply affected...intensely

troubled and fascinated by what I had just seen.

Tom Brown's review continued on back

New Videobox Artwork

As marketing efforts on PiG continue, new concept art for the video box has been developed. Designer Larry Herbert combined elements of the original promotional art with some new ideas to come up with a striking and eye-catching concept. A limited number of videos will be produced and packaged with this design for use with distribution and media representatives.

Obviously, all those who worked so hard to make this film a reality want to view the finished product and video copies of PiG will be available sometime in the near future. Editing changes, credits and final sound scoring unfortunately delayed this from happening sooner.●

The newsletter also had space for a short handwritten note, which usually detailed the film's status and any media coverage the production had received to that point. Maintaining contact with the industry was our goal; we mailed four of these updates over the course of the shoot and postproduction phase of the film. Without the budget to buy advertising or the stars to attract the attention of industry media, it was a smart and cost-effective way to personally deliver consistent news on our project to our most important target market.

Finished Product Sales

In the world of independent video and filmmaking, it's best to approach distributors with finished product. And that's more than just my humble opinion. With more than 3,000 feature films and at least an equal (but probably greater) amount of videos produced and promoted annually, it's hard enough for acquisitions agents, distributors, and other entertainment product buyers to sort through the screeners on their desks, much less consider proposals for projects that need start-up capital or completion financing. Remember, we're talking independent here, not Hollywood-level pitches for films starring "A" list talent and featuring exotic locales and leading-edge effects.

Reality always has to be part of the game plan for independent producers, as those who choose to ignore the concept will get a big dose of it when they go knocking on distributors' doors. The reality here: You are nothing special. Your mom might think you are special. The same goes for your significant others and pets. Hell, even I think you're special (you *are* reading this book). But a distributor looking for new product? Sorry, he's not going to be too impressed with a handful of ideas from a camcorder-toting wannabe-filmmaker from Nebraska. That's just life.

You see, everyone can have an idea. Not many folks, however, can see that idea through to a finished 85-minute film. And when you are capable of putting that on your resume, you'll get a lot more attention from the people on the other side of the table. You want to go into the situation with something to offer, not the other way around.

This is not to say that you should go off and make your film or video in a vacuum, then hope to re-enter the world with plans for a six-figure distribution deal. It doesn't work that way either. While the odds for success are substantially stacked in favor of those folks soliciting something more than words and ideas, unfortunately — finished product or not — there's still the chance for disappointment.

THREE-PRONGED APPROACH

Distributors have become jaded to the promises of grandeur independent filmmakers use to describe their forthcoming projects. Show them a grandiose project and you'll get a whole different response. Put simply, completed projects offering easily identifiable marketing hooks, accessible genre conventions, and clearly defined audience segments jump to the front of the line.

What's the best way to achieve this sort of response when soliciting distributors with a finished project? Try the following three-pronged approach:

1. Survey distributors as early as possible in the life of your project and inquire as to their initial interest in, and potential salability of, the subject matter.
2. Throughout the production and postproduction phase of the film or video, keep any potential buyers updated (via electronic or traditional mail) on the status of the project and any media, fan, or industry attention it receives.
3. Officially solicit distributors only when your project is as complete as you can make it. Including key art, reviews, media coverage, and marketing suggestions in the package can only sweeten the pot.

This kind of concerted effort involves and addresses potential distributor needs throughout the entire life of your project. The technique serves to provide continual information to maintain interest and awareness, while at the same time highlighting the marketable elements of your film or video, allowing you to approach buyers in a professional and prepared manner.

Chapter 11 Summary Points

■ Independent producers should contact studio and large independent distributors first to increase the chance of the widest possible release (and thus success) for their project.

■ It's tough to market a project directly to mass merchandisers and other large retailers without the help of a distributor.

■ Soliciting distributors during a film or video's development stage provides a good indication of whether the project is something the market will ultimately accept.

■ Use the production period of your project as a time to spread the word about your soon-to-be-finished film or video.

■ Serious distributor solicitations for independently produced product should take place when the film or video is very near or at the finished product stage.

PRESENTATION PACKAGE

If a distributor doesn't believe your project can make money, he's not interested. That's a pretty strong statement, which probably stings many readers. But that's life. And business. A distribution company cannot survive by being good guys and helping every independent producer that comes knocking. Sure, a distributor may elaborate on your project's beautiful cinematography, excellent acting, and great storyline, but he can't help you if you can't help him make money.

The production merits of your project — whether it's a deck-building instructional video or a low-budget action feature — count, too, and are always a determining factor in any acquisitions executive's ultimate buy or pass decision. A shoddy film or video, no matter how marketable the idea may seem, won't get picked up.

But let's say you've jumped that hurdle and created a high-quality project, one also packing a couple of marketable hooks and recognizable audience appeal. What else can you do to improve your chances and make the project more attractive to prospective buyers?

In the competitive marketplace, you need all the help you can get to stand out from the crowd. You need to distinguish yourself from the pack — make a reason for a distributor to select your film or video over others of like quality and subject matter. You've got to give these people a reason to stop saying "no" and convince them (or at least nudge them in the right direction) of the profit potential of your project. You've got to sell them. And the proper way to do this is with a comprehensive presentation package.

The presentation package is the physical culmination of all of your promotional efforts. This multi-piece marketing tool can consist of many items, some of which might include:

- Cover letter (which is your sales "pitch")
- Media kit (comprehensive data on all aspects of the project)
- Full-length screener (a complete copy of your program)

- Key art (what the video sleeve or DVD package may look like)
- Production and publicity photos (vital to most distribution deals)
- Ancillary marketing merchandise (promotional items that "advertise" your project)

Why go to all this trouble? Why not just slip a VHS screener in an envelope and hope for the best? Unfortunately, the most effective means of landing a distributor — personal selling — is unavailable to most readers. Thus, this sales "kit" must do the selling for you. While it can't prop up an ill-conceived production, a polished presentation package can add that extra incentive needed to arouse more than a cursory examination of your project by those who matter. You need to create a total "environment" for the distributor as he considers your proposal. Anything less, and you should probably settle for less. This is your one shot. Remember that when you consider taking the easy way out.

SALES LETTER

Publishers Clearing House. Time-Life Books. Columbia Record Club. All examples of companies whose long-running success can be attributed to the use of outstanding sales letters. What's their secret? Think about how you react to one of their pitches. If you're like me, you probably feel like you'd better take part in the advertised fabulous offer or you are going to miss out on something big. Somehow, the authors of these promotions are able to convey a feeling of honesty, even though the intent of the piece is to sell you something. And these solicitations consistently avoid the "scam quotient" so prevalent in many direct mail solicitations. If that wasn't enough, the letter makes it easy for you to participate in the program. What more could a potential buyer need?

Well, investigating the "feelings" generated by such skillful sales prose exposes the secret to the letters' power. A successful sales letter must do two things:

- Create a desire on the part of the reader
- Demand an action on the part of the reader

If you can satisfy these two elements — desire and action — when composing your sales letter, then you are well on your way to developing what may be the most difficult-to-create portion of your presentation package. There is, of course, more to it than that.

Director/Producer John Ervin used a quick-and-simple approach to his film's sales solicitations.

December 8, 2001

Well Meaning Independent Distributor
Attention: Acquisitions Director
XXX Street Ave, Suite XX
New York, NY XXXXX

Re: "Vixen Highway"

Dear Acquisitions Director:

At long last here is the NTSC VHS tape of "Vixen Highway" you had expressed interest in viewing and for which you have been waiting all these months. Many apologies for the long delay, but technical, financial and other matters came up that either slowed down or distracted significantly from completion of the arduous post-production process. In any event, I hope you enjoy my digital video spectacular, which runs 70 minutes total. Please feel free to pass along any questions or concerns via e-mail, phone or, if you are so enclined, snail mail. Thanks again for your consideration of the material and I hope that we can bring "Vixen Highway" to theaters, festivals, rental and retail outlets the world over!

Best Regards,

John Ervin
Berlin Productions
Phone: (612) 362-5956
e-mail: prolix@earthlink.net
www.vixenhighway.com

Hit 'Em Hard

All effective sales letters have three distinct sections: the opening, the body, the close. The opening of your letter must immediately attract the reader's attention. If you stumble right out of the gate, failing to grasp the reader's interest, she may never finish the rest of the letter. Acquisitions executives and other film and video buyers read thousands of letters a year. You are going to have to do something in the first two lines to be sure it gets read to the end.

While an opening such as "Enclosed please find..." certainly gets the job done, are you captivating the reader with this kind of writing?

Hardly. Studies show that you have about four seconds to grab the attention of your reader. So you better come out of the gate with a hot opening line. It's the most important line in your entire letter. You are up against a lot of competition for the attention of your reader. Every word, every sentence of your letter is important.

Using attention-grabbing phrases and words will usually keep the distributor interested enough to continue reading. Don't write dry, corporate drivel to open your letter. This is the entertainment business — try to excite, even shock, the reader.

Keep 'Em Reading

Creating desire for your product is the function of a sales letter's body. Ideally, you want a potential buyer to become so intrigued with the marketing possibilities inherent in your film or video, that she'll feel the need to watch the enclosed screener. You are setting her up to take some action.

The body of the letter should be used to speak of anything new the product has to offer — essentially the film or video's hooks. Even historically successful independent film genres (horror, road-movies) and video product lines (exercise, motivational) must be capable of offering an audience something new, innovative, or cutting-edge. That is what "break-out" films are all about: high concept ideas executed in such a way that the genre seems new again.

Horror films have been always been a stable of popular cinema. In the early 1980s, the look of horror films changed with two breakout movies, *Halloween* and *Friday the 13th*. While these films offered nothing new with regard to their basic storylines, the former introduced moviegoers to a unique point-of-view, and the latter pioneered never-before-seen special effects. Try to think of your movie in this way when describing it in the body of the letter. Regardless if your film or video deals with time-worn conventions, now is the chance to paint it as something different.

You want to create an impossible-to-resist first impression before the buyer has a chance to create one of her own based on the title, implied subject matter, or any other aspect associated with your project. These sometimes intangible elements may be saying something to her before she realizes it. It's called "profiling" when dealing with humans, but this practice of judging by appearances or associations is a part of sales decisions, too. The goal here is to use the sales letter to assist in the proper profiling of your product, to steer the thinking of the reader toward the benefits your film or video offers.

Actual sales letter sent to prospective distributors during the first round of solicitations for the feature film Pig

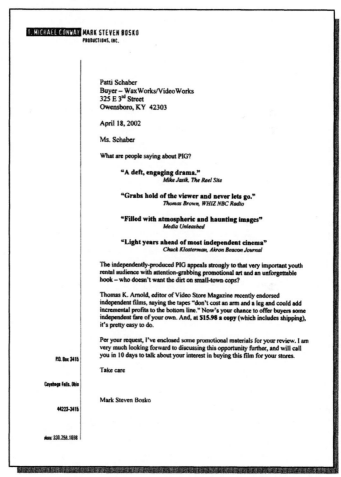

The body of the letter also gives you an opportunity to brag about your project's achievements. Good reviews, festival awards, a very active Web site, featured personalities or stars, media attention — anything that could be used to support the marketing of the film or video should be

highlighted. Don't go into excessive detail, just give enough information so the potential buyer is enticed into investigating the package further.

Finish 'Em Off

Having convinced the reader that your film or video is worth additional examination, you want to get action before she has a chance to change her mind, move onto the next solicitation, or simply forget about your project. Asking the buyer to take action is the purpose of the close, the last section of the sales letter. The letter must ask something of the buyer. Don't make the mistake of failing to ask the reader to take some kind of action on your project.

A good "clincher" — as the close of a sales letter is known in the advertising business — includes two essential parts:

- A clear statement of the action wanted
- A description of the benefit(s) gained by the action

When soliciting a buyer, the first action you want from the person reading the letter is to watch the screener. Ideally, she should finish the letter and immediately begin watching the film or video. Ask the reader to do just that. Better yet, tell her. Try closing with a sentence that starts like, "Watch the film...," plainly telling the distributor the desired action.

You also want the buyer to decide if the project is something her company wants to distribute. Hopefully, your research directed you to buyers offering the best probability of this occurring. No letter, regardless of how well it is written, is going to fully persuade a buyer. But you can at least get the distributor thinking about a decision by pointing out in the letter that you will "call in 10 days to discuss." A distributor will contact you after watching your film or video and examining the contents of your presentation package *only* if they are hot for the project. In most cases, the burden of second contact (the call after the package is received) falls to you, the solicitor.

Minneapolis-based independent film producer Todd Wardrope's sales letter for his feature, Nuada

It's vital to mention a benefit the reader will gain by fulfilling the call to action. For traditional sales letters — such as those Publisher's Clearing House solicitations — this usually takes the form of a time-sensitive special savings offer. You know, "…act now and save an additional 10% off the regular price." But when initially shopping for a distributor, this benefit statement must concentrate on the non-material aspects associated with watching your project in a favorable manner. Words that talk to the audience and profit potential of the project are good choices, as is a quick reminder that the film or video is complete and ready-to-go, requiring little investment on the buyer's part to put it into the market (this is assuming you have fulfilled all technical production requirements). Remember, you want to approach potential buyers with your hand full, not empty.

Prose Pointers

As you compose your sales letter, try these suggestions:

- Step into the buyer's shoes. Ask yourself, "If I were the buyer, what information would I need to make a decision, to immediately want to watch this film or video?" Distributors aren't interested in how you carried off a special effect, even if it is your production crew's pride and joy. They want to know why your film will sell. Why is it popular with audiences? Why will people pay to view it? Instill confidence in the buyer by acting as if your project is already a success. Focus on their needs, wants, hopes, and desires. Play to their emotions.

- Use a question, quote, shocking statement, or humor to open. All of these devices involve the reader, drawing her into the letter. She'll feel the need to read more. The first sentence is the single most important line in the entire letter.

- Always feature the positive. Always forget what could be considered negative. Think very carefully about each sentence before committing to use it to the final version. So what if your feature-length film is shot on Super 8? This could be a badge of honor to some independent filmmakers, a testament to your determination to get a film made no matter what the restrictions. Most distributors won't see it that way. Super 8 to them means grainy, means amateur, means extra cost if they have to blow it up. Don't create a negative image in the buyer's mind before she even sees the film. Stress instead the great publicity you received. Talk about the project's ability to deliver on the genre. Point out the interesting soundtrack. Think and write positive.

- Use YOU, not I. Writing from the reader's point-of-view (you-words) creates a more positive involvement with the letter. Don't use sentences like "I believe...." Write instead, "You'll agree...." The use of "I-words" alienates the reader. She must be involved to care.

- Write like you talk. Don't use unnatural language that is all too common in business communications. Sentences like, "Please take note of the enclosed press clippings," and "At the present time, my film is available for video distribution," sound lifeless. Change those sentences to "Check out the unbelievable amount of media given the film," and "You can distribute this film now," and you add a more active, genuine tone to the letter. Forget about always writing in complete sentences. A good test of your sales letter is to read it aloud. Does it sound natural? Does it sound as if you were making a pitch to the distributor in person?

- Don't be stiff. It's okay to use slang. Describe your film or video as hip, cool, hot, gotta-see, etc.

- Keep it short, but tell the whole story. A successful sales letter shouldn't exceed one page in length. There's just no way an acquisitions exec is going to read past that. Describe just enough background so the potential buyer has a good idea of what you are trying to sell. Don't get bogged down with pointless details that have no impact on the marketability of the project.

- Don't be pompous. Convince the buyer that she is important to the success of the project. Never give the impression that it's okay for the reader to pass on the film or video since so many other buyers are begging for distribution rights. This ploy will not work. Distributors are a tight-knit group of businesspeople. They know what has and hasn't been offered to each other, what's good and what's not.

- Use the reader's name. You need to write to someone specific. Try to imagine the individual receiving the letter. Make her a live, breathing person, not just another name on a list. Refer to the buyer by name when calling for action. This brings the request to a more personal level.

- Sleep on it. A letter that sounds great today may not tomorrow. After composing your letter, put it aside for a day or two then reread it. You'll be surprised at all the room for improvement. Never use a first draft. Good writing is really all about rewriting.

For those who haven't done much business writing, the mechanics of a sales letter are pretty simple. Include the date and the buyer's name and address at the top of the page. Single space within paragraphs, double space between them.

MEDIA KITS

Though primarily used for enticing the media to cover your project (thus the name), a media kit is just as valuable for soliciting distributors. This comprehensive package provides detailed information on all aspects of your film or video. Built as a collection of parts, the media kit acts as a salesperson in your absence. Giving you the opportunity to go beyond a simple cover letter taped to the front of a VHS screening tape, a well-conceived and executed media packet tells the whole story, enhancing the project's image and giving your presentation extra "oomph."

Media kits also serve as a good "foot-in-the-door" introduction to the inevitable follow-up calls you'll be making. Prospects receiving exhaustive kits respect the effort and the fact that you approached the sales process in a professional manner, and will usually respond to your inquiries in like manner.

With program length, format, cast, crew, location, story, and contact information at the recipient's fingertips, a good media kit saves time by eliminating the hassle of follow-up phone calls to verify any initial inquiries of this nature. After mailing the kits for *Pig*, we received several compliments from prospective buyers as to the thoroughness of the data provided. Most distributors could determine pretty quickly the prospective audience, genre, and marketability of the film — all from the contents of our package.

Be warned that just as a good media package helps your efforts, so does a poorly executed kit hamper your potential. Be sure the kit and its contents present your film or video in a complimentary and attractive manner. Not just aesthetically — being sure it all looks nice — but for marketability as well. Though you might be proud of the fact that your drama shatters the conventions of the genre, and want to write about it at length (something better left to a recognized reviewer), this approach may not appeal to a distributor looking to please audiences not tolerant of such change.

Creating and mailing a media kit is not cheap. The multiple printed pieces, photography, duplication, and assembly — not to mention the time it takes to design, gather, write, and edit the information, plus postage — add up. Don't let budget constraints alone put an end to this important promotional tool. The benefits of such a device are obvious, as is the fact that you'll need this packet later, if your initial distributor search doesn't land you a deal. At that point, the kit will be used to garner press as well as to inform second- and third-tier buyers as to the merits of your project.

Putting It All Together
A media kit usually consists of the following:

- Double-pocket portfolio
- Film or video synopsis
- Project fact sheet
- Cast and crew lists
- Company backgrounder
- Production photographs
- Key art
- Business card

Let's take a look at each component and how it helps to build an impressive image of your film or video production.

Double-pocket portfolio – This holder allows you to include all the elements in one easy-to-present, organized piece. You'll want to choose a folder that complements your production company's stationery design and colors, or the artwork/photography associated with the film or video. For both *Killer Nerd* and *Pig*, the key art we designed during preproduction was color-printed onto label paper that was then affixed to the front of glossy black pocket-folders (available at any office supply store). It looked very professional, and cost about $1/10^{th}$ of the price of custom-printed folders.

Film or video synopsis – A summary of your project, told in story format. When you are using the media kit to solicit the media, you'll want to include two of these: a long form version that depicts the movie or video in plot-point-by-plot-point detail and a short single-page description of the main action. For distributors, the short version will usually do. They're not reporters. They don't need all the details in print if they have the screener in their VCR. The synopsis should be a true representation of the product, not a hype-filled vision of what you would like people to believe. Many distributor Web sites contain a product synopsis section. These provide perfect examples for what to include in your kit. Surf several to get the feel, tone, and voice of the writing.

Project fact sheet – Consider the questions you would have if acquiring a film or video from an outside source. You would want to know the genre of the project and the projected target audience this film or video was aimed toward (including ages, sex, occupation, and education level). You would want information on the length of the production, the film or tape format it was produced with, and any technical specifications such as Dolby encoding, aspect ratio, and the availability of separate music and effect audio tracks. Finally, you may want to know project details such as the source of the story, if a video is part of a series, where the production was created, and who is playing on the soundtrack. Laying out all this information (and any other facts you may believe applicable) in an easy-to-read style will allow your prospects to view, at a glance, the pertinent information they need to make a decision.

Dragon and the Hawk Data File

Dragon and the Hawk Data File

Movie Title: Dragon and the Hawk
Year: 2000
Theatrical Release Date: March 10, 2000
Soundtrack Release Date: October 31, 2000
Video Release Date: October 31, 2001
Studio: Inferno Film Productions, LLC
Manufacturer: Inferno Film Productions, LLC
Distributors: Westlake Entertainment Group; Inferno Film Productions, LLC
Director: Mark Steven Grove
Producer: Darlene Cypser
Screenwriter: Robert Gosnell
Starring Actors: Julian Jung Lee; Barbara Gehring; Trygve Lode

MPAA Rating: R
Genre: Action / Adventure
Keywords: Action, Detectives, Martial Arts, Mystery

VHS Technical Details: NTSC; Color, Full Screen (Pan & Scan) 1.33:1; Dolby Surround (Prologic); Runtime: 87 minutes; Languages: English; Closed Captioned: No; MSRP: $9.99; UPC: 803236100236

DVD Technical Details: Region 0 (All Regions); Color; Full Screen (Pan & Scan) 1.33:1; 2 soundtracks: Dolby Digital 5.1 and Dolby Surround; Feature Runtime: 87 minutes; Special Features: Animated Menus, Trailers, Music Video; Number of Disks: 1; Single layer; Single Side; Languages: English; Closed Caption: No; Subtitled: No; MSRP: $14.99; UPC: 803236100298

Soundtrack CD Technical Details: Number of Disks: 1; Number of Tracks: 24; MSRP: $10.99; UPC: 803226100120

Brief Description 1 (60 words): Dragon and the Hawk is an action/adventure film that draws the viewer from the pedestrian daylight world into a dark underworld of crime and human experimentation. In a style somewhere between Xena and Lethal Weapon, Dragon and Hawk fight their way from the city streets through the dungeons of the Inferno to the laboratory of the sinister Therion.

Brief Description 2 (71 words): Martial arts master Dragon Pak (Julian Jung Lee) is the man who will stop at nothing to find his missing sister. Lieutenant Dana "Hawk" Hawkins (Barbara Gehring) is the cop who risks her career...and her life... to help him. Together they plunge into the dark and sinister world of the evil Therion (Trygve Lode)...a criminal underworld of hidden laboratories, women chained in dungeons, and human experimentation in pursuit of world domination....

A one-page fact sheet provides all pertinent acquisitions information for those interested in the independent film, Dragon and the Hawk.

Cast and crew lists – Beginning with the lead of a film or the host of a video, list every on-screen participant, their role, and any other known or appropriate past experience. Do the same for crew, including writers, producers, and directors, in addition to the traditional camera, lights, set, wardrobe, and make-up people. Point out special accomplishments, awards, and training for those in each group.

Company backgrounder – This document should include a brief history of your company along with any recent or past business news. Don't make it a brag sheet, rather use it to paint a professional image of your company as one able to create marketable products for the entertainment industry. Tell of past successful productions and explain in active language why your company is tuned into whatever film or video genre you are exploiting. The profile should be approximately one page in length.

Production photographs – The old saying, "a picture is worth a thousand words," never holds truer than when you are discussing busy acquisitions agents. Though the rest of the materials in the media kit are appreciated, your photos and key art (and later, packaging designs and marketing materials) are what will grab their attention. Use a couple of color 5"x 7" glossy shots of a good action scene, special effect, or interesting wardrobe, set, or location. Hopefully you were forward-thinking enough during production to realize the need for these kinds of pictures. Behind-the-scenes photos are fun, but worthless except for some possible media use. It's the production shots, those pictures that look like a frame from your film or video, that count. Beyond the media kit, you'll need plenty of good, quality production photographs to close any distribution deal. So if you haven't yet shot your project — keep this need in mind.

Quality production photos, such as these from the feature film Reversal, *are an integral part to any presentation package and a must for increased media coverage.*

Key art – This should represent the project's single most promotable element. For example, the key art for *Killer Nerd* was an image of the nerd in front of a blood-spattered wall, along with the tagline, "This nerd is <u>really</u> out for revenge." *Pig* was represented by a photo of the lone police officer, standing next to his cruiser in the middle of a deserted country highway. Both images quickly "told" the story of their respective films. Single images work best, as a cluster or grouping of images is confusing and distracts attention from any one theme. Key art can be a photo, graphic, or text, depending on what type of project you are promoting. A video on public speaking might be best represented by a clever design of those words, while a martial arts feature will best be served by an action shot of the starring kung-fu warriors. Note the trend in Hollywood of using images of the stars as key art for many major motion picture campaigns. These "floating-head" designs sell the movie primarily on the strength of the audience's attraction to the project's leading actors and actresses.

Consider the font, size, and color of the title or any logo that you develop in concert with the key art. Color is also important when considering both the genre and the intended audience of the production. A horror film probably shouldn't have much pink in its key art design, nor would any do-it-yourself tape.

Developing and producing key art allows even the most cash-strapped producer to promote his project in a manner consistent with the big studios. All you need is a competent artist and some good ideas. Thanks to desktop publishing and the proliferation of graphic designers, professional-quality key art can be had for less than $500 (I've bartered art fees in exchange for advertising the artist's name on all promotional materials). Freelance artists are everywhere, and will be glad for the work if you promise to throw their name around and give them a copy of the finished product. Most designers have little experience in film or video key art design; just give them some examples of designs from video packages, posters, or magazine ads that appeal to you and seem to service an audience like yours. Though studio and large independent distributors will develop their own campaigns, regardless of how slick your design may be, this exercise helps "brand" your project and allows you to develop a consistent look across all your promotional materials. Don't cheap out here and do it yourself, unless you are a trained designer.

Business card – Even if your contact information can be found on other pieces within the kit, it is customary to attach a traditional business card. Including a business card is a simple courtesy that prevents the potential buyer from having to search for your phone, fax, e-mail, or Web site information.

The business card is part of what is known as a **stationery package**, something that you will want to develop before communicating in any way with anyone. In addition to a card, you'll want letterhead, business size envelopes with a printed return address, mailing labels, and possibly note cards. Again, use the services of a low-cost, freelance graphic designer for this work. Another option: quick-copy shops offering these packages with template designs. Kinko's is well-known for providing a huge array of low-cost business stationery items. The only downside is that your ability to customize the design is limited. All of these pieces help to "sell" the image of the project; their impact should not be overlooked. Think about the feelings you get when opening a letter printed on good stationery with a striking logo design versus a cheap, photocopied direct mail piece in a plain white envelope. Which do you give more attention? Which do you take more seriously?

Creating a professional media kit is vital. Unfortunately, distributors are subject to far too many "cheap" looking packages. Either simply printed on standard copier paper, or worse yet, photocopies of originals, these amateurish attempts are then stapled or paper clipped together and shoved into an envelope. That approach won't help to create the ideal environment you want for a distributor screening, nor will it help in getting you any serious consideration. Print all your materials as originals onto laser or better quality paper. Have art printed on glossy or photo-grade stock. Place all items of the media kit neatly in a heavyweight, pocketed folder.

Remember, prospective buyers don't know you. They are time-crunched people with more than a couple of other film and video proposals stacked on their desks. In trying to make a decision on your project's merits (and even deciding if it warrants a view), the only thing they have to go on is what you offer them. Regardless of the quality and true marketability of your film, first impressions mean everything. Make sure professionalism is a primary consideration in everything you create and use in your presentation efforts.

SCREENER

It's all about the product. Your independent feature film, documentary, or instructional video must offer the potential buyer some kind of exploitable marketing hook. There has to be something within the contents of your program that makes a distributor take notice and believe he can sell it to an audience. Otherwise, he's not interested. And the only way he can make that decision is to see what you've got.

That's where the screener comes into play. Presented to any potential distributor, buyer, or media representative, a screener exhibits how you assembled talent, writing, effects, camera work, and budget into a final product you hope they find worthy of their distribution, sales, or promotion activities. Now is the time for your film or video project to live up to its implied potential.

To save money, some independent producers will send a short, 3-5 minute trailer first, allowing the distributor to gauge — in a more concrete manner — if the film or video is something his company wants to pursue as an acquisition. For example, even though you may have described your project as an ode to the science-fiction epics of the 1950s, the intrigued distributor still isn't clear on what that means. Are you using floppy cardboard sets and rubber space monsters, or have you put a modern spin on an old style, using state-of-the art digital effects to mimic the technological failings of that era? Only by actually seeing some footage, can he be sure.

While the method is cost-effective (as duplication pricing for trailers is a fraction of what full-length program copies cost) and can save a lot of time for both parties if there is some question as to content, genre, or look of the product, I would recommend against it and suggest taking this route only if the distributor suggests. (I should mention that, though uncommon, I have dealt with some buyers who will ask for a trailer in an effort to cut their screening time. If the project looks promising, you will then be asked to submit the entire presentation package.)

Always send those distributors that you are interested in seriously soliciting full-length screeners. Potential buyers must be able to view the entire film before they decide to do business with you. Distributors know, like you do, that a good trailer doesn't necessarily guarantee a good film. Those exciting action scenes in the trailer edited together in firecracker fashion may actually take place 30 minutes apart in the finished film. Using a trailer in the presentation package to distributors looks unprofessional and may mistakenly lead to the assumption that you are seeking financing to finish the film (this happened to me). Some suggestions:

Use Only High-Grade Tapes

For solicitation purposes, duplicate screeners on the best quality tape you can find, or burn DVD copies. It is not wise to try and save a few bucks at this point by creating cheap, inferior-quality screening dubs. Nothing will cause a distributor to lose interest quicker than a tape filled with dropout and static. This is your one shot to look your best. At the same time, be sure you are working with a quality duplicator. Ask for samples of their work, or, better yet, be on-site for a supervised session. This will permit you to tweak colors, sound levels, and image quality of the dubs as they're "running." Leaving the job in the hands of a facility technician, regardless of his expertise, doesn't always ensure the best results.

Create Professional Labels, Boxes, or Packaging

While you don't need to produce a four-color video sleeve or DVD case just yet, you should try to create a professional-looking package. Print any logo on face and spine labels. If a logo has not been developed, use a quick-copy center to create high-quality labels that simply feature the program name. Blank video sleeves and DVD casings are available in a variety of colors, designs, and finishes (Markertek Industries, *www.markertek.com*, is one low-cost source). Be sure that the project title, your name, contact phone, and e-mail are clearly marked on both the exterior package and the media you are utilizing. Screeners and presentation packages have a mysterious way of separating from each other — especially in an over-solicited distributor's busy (and messy) office.

Mark Its Ownership

Print a copyright notice on the program and any packaging.

Limit Duplication

The number of first-tier potential buyers produced through your distributor survey may not be very large. Create only enough screeners for this group, along with 3-5 extras. Uninterested parties will return the tape or DVD within a couple of weeks. These can be repackaged and mailed back out to other distributors on your list. This money-saving practice is another reason to use high-grade tape and quality duplicating services.

Protect Your Property

In the duplication process, superimpose the words "For Promotional Use Only" throughout the program. This service is common, and usually free, with most duplicators. Why? Unfortunately, there are some less-than-reputable firms operating in the industry, and if you forward them a clean copy of your film or video, you never know where it may end up. I'm not inferring that distributors are dishonest, but it's always better to play it safe.

Chapter 12 Summary Points

- Quality presentation packages add clout to an independent film or video solicitation, and could include a sales letter, press kit, trailer/screener, promotional art, and photos.

- Successful sales letters create desire and demand action from the reader.

- Use an attention-grabbing phrase or quote in the opening of a sales letter to immediately catch the reader's interest.

- The body of a sales letter is a good place to make your film or video appear hard-to-resist by describing the project's hooks and market potential.

- One of the best methods for writing an effective sales letter is to place yourself in the buyer's shoes, considering his needs, wants, and desires — not your own.

- A media kit serves as a "salesperson" in your absence — comprehensively providing detailed information on program length, format, cast, crew, location, and storyline.

- Just as a good presentation package can facilitate your relations with distributors, a poorly executed one can place you at a disadvantage.

- Production photographs are an important item often overlooked by independent producers.

- Developing professional and representative key art allows for a low-budget independent production to be promoted in a manner consistent with studio-financed product.

- Include high-quality screeners (full-length copies of your project) in presentation packages for those distributors you view as serious prospect solicitations.

ORDER OF CONTACT

You already know that you should seek out the studio and large independent distributors first. These guys offer your best chance of a widespread release for your project should they decide to acquire it. Players of this size have their fingers in every distribution nook and cranny, from theatrical and home video to foreign markets and the Internet. A "major" will either find or create a market for your product, and do so in a turnkey operation. As long as you can come up with the deliverables — those items specified in the contract such as a certain number of production photos, proper audio tracks, and clearances from all talent and locations involved — your film or video will be on screens and shelves everywhere. That's the distributor's end of the deal, and that's why working with these kinds of companies is so desirable. They handle all the leg work. You maintain your designation as an independent filmmaker and can begin preparing your second project.

Obviously, this is the proverbial best-case scenario, and the odds of this happening are slim. Someday, you'll probably become a film promoter (in addition to maker), and find yourself working just as hard getting your film sold and distributed as it was to produce. But for now, you have to take this shot and believe that one of these major players can lead you to that elusive success so sought after in the entertainment industry.

In the first distributor solicitation for *Pig*, we identified the following companies as first-tier buyers — those firms that, following phone conversations with the principals involved, we believed would be most interested in the project while at the same time offering us the greatest opportunity for a large-scale release:

- Artisan Entertainment
- Columbia TriStar Home Video
- Fox Searchlight
- MGM/United Artists
- New Line Home Video
- Paramount Home Entertainment
- USA Films
- Warner Home Video

These are the home entertainment divisions of the parent companies, our targets since *Pig* was created as a direct-to-video feature. We cut and mastered the film (which was shot in 16mm) on digital video, but in a way that would allow us to go back to the original reels if a film print was ever needed.

While another 5-7 large independent distributors fell into the same category, the plan was to approach the studios first to exhaust all possibilities with that group of buyers. Not all projects are applicable for solicitation to studio buyers. Our film, however, fit many of these companies' requirements for release, so this approach was taken. We didn't approach the theatrical divisions because the costs involved in blowing up a 16mm original to 35mm for projection, couldn't be justified by the limited-appeal subject matter of *Pig*. No distributor would take on that kind of burden for a project unless it could be marketed "wide." Do not waste an acquisitions agent's time by soliciting a film or video that in no way, shape, or form resembles the type of product which they regularly release into the market.

THE MAILING

Before mailing any package, you'll want to call each company on your first-strike list to verify contact information, and let the person to whom the package is addressed know that it is coming. This simple courtesy saves you time and money on re-mailings. Believe me, there's nothing worse than mailing a screener and media kit to a distributor, only to find it two weeks later back in your own mailbox, stamped "address unknown." And then there's the issue of the packet getting directed to the wrong individual within the organization. Make the call and double-check all details. It will only take a moment and will save you from an embarrassing situation when you place the follow-up call.

If you are given the opportunity, speak with the buyer herself during this call. Most often, you'll confirm all the info with an administrative assistant, but ask to speak with the person you are soliciting. This will afford you another chance to "pre-sell" your project and alert the individual directly of its impending arrival. After rechecking all your data, keep an eye on the following:

Use a padded mailer – You don't want your screener to arrive on the prospective buyer's desk crushed from mail handling because you saved a buck on the envelope. Use well-padded bubble or foam envelopes, or mailers specifically designed for tape and/or DVD shipping. These cost more than an envelope, but ensure safe transport of your materials.

Use priority mail – You don't need to utilize an overnight delivery system (unless so instructed by the buyer), but you also don't want to ship packages via standard mail. The U.S. postal system's Priority service gets packages anywhere in the country in about three days for less than four dollars. Priority mail is handled better, and just looks a lot more serious and professional than standard delivery.

Use easy-to-read labeling – Don't use fancy, small, or hard-to-read fonts. Simple is best. Place the recipient's name above the company name and address in clear, bold, black lettering. Don't forget to include return address information on all packages.

Make sure all contents are in package – In your excitement to start the solicitation process, don't forget to include all items in each kit. Check for cover letters, screeners, and all parts of the media kit before sealing any envelope. Also check that the name on the cover letter matches the name on the address label. Mixing up mailings can be an easy mistake when working with more than 10 packages at once.

Be appropriate with creativity – As a film or videomaker, it is your job to be creative. But should this apply to the presentation package mailing? That's pretty much a judgment call. When soliciting distributors for *Killer Nerd*, we mailed the presentation packages in acid-green envelopes that featured the film's logo on the back. On the front of the package was a small sticker containing the film's tagline, "This nerd's really out for revenge." It was appropriate for the tongue-in-cheek humor of the horror-comedy, and many of the package's recipients commented on its attention-getting quality. For *Pig*, we used plain black envelopes, and again placed the logo on the back of the mailer. There's a fine line between a professional, unique package and one that looks like a high school art class project. Carefully consider any creative flourishes you think will enhance the packet. If well done, and in line with the theme of your project, the extra work will at least get your package moved to the top of the mail pile. It's just one more way of creating a favorable environment for the delivery of your materials, and also puts you a notch above any other proposals coming in that day. The use of creative mailers won't guarantee a positive response to your solicitation, but they won't hurt your chances, either.

THE FOLLOW-UP

If you've started the solicitation process, the last couple of weeks have probably been pretty hectic for you with all the phone calls, writing, preparing and assembling screeners and other media kit materials. Now that the presentation packages are in the mail, take a moment for a well-deserved break. I'm not saying to forget about the project entirely, but try to involve yourself with non-film or videomaking pursuits for a bit. Beyond waiting until your first attempt at callbacks, there's not much to do except worry about the response you'll receive from those solicited. So don't obsess (at least not too much) on what could or couldn't happen. Your fates are, as they say, with the Gods. It just so happens the Gods in this scenario run distribution companies.

THE FIRST CALL

Wait at least two weeks from the day you mailed the packages until you start calling distribution companies to discuss their interest in your project. Why so long? First, it can take up to five days for delivery, even with priority postage, as packets mailed on a Friday may not get processed by the U.S. Postal Service until Monday. Then, depending on the size of the company, there can be some mailroom processing time. Add in a weekend or two when the acquisitions executive is not in her office, and the fact that she may have 5-10 or more other solicitations to sort through, in addition to her other job responsibilities, and several weeks seems like nothing. But it is enough time for most people to actually receive the package (if it was addressed correctly), and that's often all the first call accomplishes — acknowledgement of receipt.

Before dialing anyone, relax. It's not like you are a telemarketer trying to sell something to someone who may have little-to-no interest in your product. You did your homework, and now is when that effort pays off. The distributor is expecting your solicitation and your call, so "just do it." When calling, make sure you are on a good phone in a private environment. Also, have a copy of the presentation package as well as a notepad and pen within reach.

If you've created an impressive and professional presentation package, you've got a jump on most of the competition. When soliciting *Killer Nerd* and *Pig*, many buyers remarked on the quality of the package, and more importantly, the comprehensive and appropriate level of information included. They liked the fact that we researched the marketplace, offering them not only a project that fit within their distribution standards, but also a wealth of support information in the media kit that answered most of their preliminary questions.

Kat Candler used reviews, art, postcards, fact sheets, and a screener as part of her presentation package.

Some distribution agents shared horror stories of describing unsuitable submissions (my favorite: a non-dialogue video aimed at toddlers called *Walking is Fun*, mailed to a distributor that specialized in ultra-low budget horror films) and amateurish media kits. Presentation means a lot in this business, and if you did the job right, you can look forward to a positive conversation.

Once connected with the specific person you solicited, introduce yourself and inquire as to the status of your solicitation. Has it arrived? Has he had a chance to review the contents? Don't get depressed if the distributor is unfamiliar with your project and the answer to both questions is "no." These are truly busy folks. I've had conversations with acquisitions agents who reserve one day a week just to watch screeners. It is not uncommon if your project has not been reviewed within two weeks of shipping. The last thing you want to do is rush anyone. Mention the fact that you were simply confirming receipt of the package, at which point the distributor will likely scan the contents of his desk (or a mail area) and locate the goods. To be honest, he probably has looked at it and (this is gonna hurt) doesn't remember. That's fine. It happens all the time. The point was to make sure your package got where it was intended to go.

That's it. The first call is over and you are back to where you started two weeks ago — left wondering. Get used to it. If you are the impatient sort, this period of your project's "life" is gonna be tough. You are just one of hundreds of promising film and videomakers looking to land a distribution contract. Waiting for an answer will occupy the largest portion of your time in this initial distribution solicitation. It's not that you aren't important, it's just that there are others in line ahead of you. Don't get discouraged, just clear your thoughts and call the next distribution company on your list.

INEVITABLE OUTCOMES
Return calls obviously don't always end with more waiting. What if the distributor did receive the package and did review the screener? What happens now? I'd like to say that this is when the gates to Hollywood open, the Porsche dealer accepts your credit application, and a stylist is waiting to make-over your look for a guest appearance on Letterman,

but the world of independent filmmaking works a little slower than that. Distributors who have reviewed your package can only offer one of three responses: yes, no, or maybe.

Rejection, while never fun, is a fact of life for people who take risks. And producing and soliciting independent films and videos is a pretty risky venture — risky for two reasons: There's a lot of money and time riding on whether you attain success or not, and you are asking someone to react positively to something you have created. Ask any artist or creative person who has put their work on display for others to judge. It is an often discouraging process. Soliciting distributors is no different. Not everyone is going to like your film or video. What can you do?

WHEN THEY SAY "NO"

Besides sulking when an acquisitions person at a distributor tells you that your project is not something he wants to work with, use the opportunity to learn. Why didn't he like your project? Was it the acting? The production quality? The format? Maybe the soundtrack? Find out what made it unattractive to him from a business standpoint, and, if he is willing and the conversation is not rushed, ask what he did or didn't like from a personal point of view. (For the record, use of non-professional talent and the inability to meet genre conventions are the most popular responses.) This kind of feedback is valuable because the people who are talking work within the market daily.

Distributors are keenly aware of what they can and cannot sell, and can immediately spot either of theses traits fairly quickly. Heed their advice. If someone tells you that a film concerning a small-town cop is going to be a tough sell unless certain hooks are primarily promoted, you probably should listen to any ideas he has on how to market the project. (Yes, that's a real-life experience.)

If possible, get information on other companies that may be interested in your now rejected project. Ask the distributor if he can direct you to someone else in the industry that may be better suited for the type of film or video you are offering.

Something unmarketable to one distributor may likely become a very profitable product for the right company exploiting a specialized niche.

This is how we made the deal on *Killer Nerd*. In initial solicitations, the feature was turned down by a number of distributors. During one of these "rejection" conversations, a distributor told me that he knew of a company in Fort Lauderdale, Florida, that was looking for new, horror-themed features. The buyer on the line didn't have any contact information available at that moment, but told me to call back in a couple of days and he'd locate some information by then. As promised, two days later he gave me the name and number of Hollywood Home Entertainment, and the rest, as they say, is history. I began the whole process again with Hollywood Home Entertainment, calling to survey their needs, preparing and shipping a presentation package, and making a follow-up call — only this time getting a positive reaction to the project. Numerous subsequent phone calls and a couple of face-to-face meetings ultimately led to their status as the company responsible for the home video release of the independent feature film.

If, after several calls, you still haven't found success, try to remember that distributors specialize, and it takes some time to find the right one. Keep at it. Also realize the sheer number of films and videos these folks screen on a weekly basis is phenomenal. Unfortunately, acquisitions agents and staffers look at a lot of bad stuff, and sometimes it's the timing that can kill you. Could you imagine not being a jaded viewer after sifting through 10-12 tapes in the course of a morning?

It's not the early 80s anymore, when pretty much anything recorded on tape could be packaged and sold. (Does anyone remember the video-tape fireplace or aquarium?) Independent film and videomakers have access to very sophisticated equipment capable of creating very sophisticated projects. But having the former doesn't always guarantee the latter, and that is what clogs up the acquisitions process — not to mention a lack of education and understanding on the part of the producer. Those film and videomakers using the shotgun approach to solicitations (sending everything to everyone) burden the system and make it harder for smaller, yet quality projects to stand out. Unless a star is attached, there are extraordinary special effects and locations, or the project features an exemplary marketing hook, it can be tough to get noticed. Accept that fact before you begin the process.

Don't forget to send thank-you notes even to those people who rejected your project. Buyers are like elephants and remember everything. Courteous and professional treatment will never do you any harm. Plus, you never know whom you, or your project, will bump into. Let's say you unsuccessfully solicited the buyer at Warner Home Video. What you don't know is that he's good friends with a buyer at Spectrum Releasing, a smaller distributor, better suited to your needs. If you were polite and businesslike with the guy at Warner, you may find yourself fielding a call from his buddy. I know that distributors talk amongst themselves about the outrageous, horrendous, and incredibly stupid things that independent film and videomakers do in their quest to land a distribution contract. But they also talk about the good sometimes, too. And you always want to be on this side of the conversation. Send the thank-you note no matter how much you want to stick pins in the acquisitions agent's eyes.

SOLICITATION LIMBO

What's worse than hearing "no" from a potential buyer? How about, "I have to think about it." Initially that answer must seem a lot better than a rejection. And, in many cases, it is. Especially when it leads to a distribution deal. The down side of this response, however, means you start playing the waiting game. Imagine for a moment that you've solicited ten first-tier companies. From that group, eight have passed on the project, with the remaining two "needing more time." At this point, you are two months into the solicitation process and you want to move on to your second group of buyers. But you can't start that second round of mailings until you've exhausted all the possibilities from the first group. And those two companies that need more time are holding you back. Now do you understand why this can be an even more aggravating response than a simple "no"?

If time is not a factor, then it's not a big deal to wait out a distributor's decision. Most film and videomakers don't want to spend more time selling projects than making them, so this situation causes a lot of frustration. The worse part — and I know from personal experience — is holding out for an answer only to hear the dreaded "no." When selling *Pig,* our first round of solicitations found us waiting for an answer from a well-established distributor. The whole time I let my imagination get the best of me, envisioning the life my partner and I would be living after signing a career-making contract. Well, after two long weeks, I called the buyer back, and he still wasn't sure what he wanted to do. He needed more time. Another two weeks, another round of fantasizing, and still no response. The guy had our film for more than a month! Finally, after a total of seven-and-a-half weeks I got the screener back in the mail with a note attached explaining that the project wasn't something they could distribute. It seems almost humorous now, and that's the reason I'm sharing the story with you. You'll need this kind of perspective to maintain your sanity throughout the process.

A non-committal acquisitions representative may also be facing the problem of authority — meaning, he can't give the thumbs-up without the blessing of others. People in this position often have a chain of superiors they need to convince as well. Just as everyone wants to be connected to a successful venture, nobody wants to be associated with a flop. And if the buyer or his superiors leave or change positions within the company, projects don't always get inherited by new hires. Again, the responsibility for selecting hits and misses is judged as something that can either make or break a career, and nobody is willing to stake his future on a deal that was put together without his involvement. Bottom line, if your project is meeting with favorable acceptance by someone, and he takes another job before you sign a contract, your film or video will most likely need a new home, too.

Another time you'll encounter a "let-me-think-about-it" response is when you've solicited a buyer who is just too nice to reject you. These guys love film and videomakers and never quite seem able to just say no. You'll discuss possible markets, they'll ask you for more information, more screeners and press kits, tell you to call back in a couple of weeks... but nothing really ever happens. This individual may truly believe he can help, but in the end, all you'll have accomplished is wasting time. If you feel this sort of relationship developing with a buyer, set a deadline for when you need a firm answer. Just be honest with the distributor and explain that you have other solicitations to make if he can't help out. If you get no response by the deadline, move on.

GETTING TO "YES!"

Guess what folks? There are times when you call a distributor after mailing a presentation package and they like your project — so much so that they want to start negotiating with you about acquiring it for distribution. This response probably sounds too good to be true, and in all honesty, it happens a lot less than one would hope. But, it does happen. And boy, it'll be one of the best phone calls you ever make!

Actually, if someone you solicited loves your project, they'll probably contact you before you get the chance to contact them. It all goes back to the theory that everyone is searching for the next big thing, and if your film or video is perceived in that manner, then a distributor wants to tie you up before someone else can get to you. The psychology of sales screams not to do this, as prices rise in accordance with demand (meaning you may ask for more cash if you feel a buyer is hot for your project), but some distributors can't help themselves.

Either way, an interested buyer is an interested buyer, and from that point on, you'll proceed according to his directions. Most often a series of face-to-face meetings are arranged, or, if geography is an issue or the project will be marketed to a niche audience, most of the dealings will take place via conference calls, e-mail, and fax (to save costs).

Chapter 13 Summary Points

■ Be realistic and approach only those distributors that handle the kind of project you are offering.

■ Mailings of presentation packages should take place only after the distributor has been called to verify contact information and made aware a shipment is on its way.

■ Remember to verify contents and clearly label every package mailed.

■ Don't get nervous when placing the follow-up call — distributors want to talk to you.

■ Use rejections as a chance to learn — ask uninterested distributors why they didn't like your project and who in the industry might help.

■ Sometimes a "maybe" is worse than a "no" if it keeps your project away from other distributors who may display serious interest in acquiring it.

THE BEGINNINGS
OF A DEAL

Regardless if you are talking about a feature film or an instructional video, initial discussions are all about "feeling" the other party out. You want to know what the distributor can do for you and your project; they want to know if you'll be able to hold up your end of the deal. What this means is a distributor that says, "I'm interested in your project, let's talk," isn't offering you a distribution deal. You're a hell of a lot closer to one than you were two minutes before the phone call, but you're not there yet. With that in mind, some cautions to avoid:

DON'T TELL THE WORLD
Sure, you can get excited. But don't go nuts. When soliciting *Pig*, we were getting close to a deal with Spectrum Films. They liked the film, identified a market, and we talked about a contract. This distributor wanted some visual and audio content changes, which was fine with us. We wanted an advertising commitment and a solid per tape price (it was going to be a direct-to-video deal), but were excited at the fact that the film would receive wide distribution from Spectrum. After all of our hard work, we couldn't help but "leak" the news of the imminent deal to several Web and print media. Well, as soon as the news hit the Internet, I got a non-too-friendly fax from Spectrum asking us to stop promoting what hadn't yet happened. While that didn't kill the deal, it didn't help our side of the negotiations. It's okay to tell friends, family, and those close to the project. Just don't call the media yet.

DON'T FORWARD ANY MATERIALS
Most reputable distributors won't ask for anything until the contract is signed. There are, however, some companies that will ask for art and high-quality masters of the project to inspect before putting all their terms in writing. This approach isn't the way the industry works. Never send anything when you are just discussing possible deals.

DON'T BE IMPATIENT

I know it's been a long trip and you are in a hurry to become famous, but don't rush the buyer during this "get-to-know-you" period. Most likely, you are not the only film or videomaker she's working with. People like to do business with people they like, especially in a long-term arrangement like distribution. Both she and you want to trust each other, and if she tells you it'll be a couple of weeks until she can call you, don't call her at 8:00 a.m., 15 days later, wondering why she hasn't called. You've waited this long, don't blow it now by goading this person toward a decision. Chances are it'll be one that you don't like.

Even though you've done some research on all the companies solicited, at this point, you may want to dig a little deeper. Check out other films or videos this company has released, and answer the following:

- Do they look like quality products?
- Are you familiar with the projects' marketing materials (packaging, posters, advertising)?
- Have they received good reviews?
- Are the company's films or videos well respected in the market?
- Do their products sell or rent well?
- Are they afforded a lot of shelf space?
- Is the company a brand name in the business?
- Do their films or videos get publicity?
- Is the company a niche marketer?

Any information collected in this step will help you further analyze a potential suitor, sizing up its capabilities and potential in providing the level of distribution, exposure, and sales you deem necessary.

If further research produces encouraging results and early conversations are positive (on both sides), you'll likely proceed toward more serious contract discussions and even begin the negotiating process.

Chapter 14 Summary Points

- Though tempting, don't brag about a possible distribution deal to media and other industry insiders until it is a sure thing.

- Refrain from sending any reproducible materials to a distributor until a legal contract is signed by both parties.

- Check out any potential buyers, paying special attention to how they treat other producers that have worked with them.

UNDERSTANDING REALITIES

As you solicit traditional studio and large independent distributors for your project, it's a good idea to keep this figure in your head:

Approximately 10-20 independent films and videos are finished daily — every day of the year.

Attached to each of these is someone just like yourself who has dreams and aspirations of becoming famous or rich or self-supporting through his passion for making entertainment, and who also believes that his project is the next big thing. But not everyone is going to get offered a deal. It is both mathematically and economically impossible. And it's not always quality that is the determining factor in the buy or pass decision.

It may not seem "fair," but four areas — that are only quasi in your control — can impact whether or not you land a distribution deal at this point:

TIMING
When you solicit a distributor is important. Your golf instructional video featuring a PGA instructor, beautiful Hawaiian locale, and "5-minutes-to-longer-drives" marketing hook appears to have all the bases covered. What you didn't know, however, is that eight of the 10 companies you've been soliciting have:

- Another video in the pipeline dealing with the same subject
- Quit selling golf videos
- Hit on hard financial times
- Begun producing all of their videos in-house
- A marketing partnership with a competing golf ball and club manufacturer than the ones featured in your video
- Or any one of a hundred other bad-timing issues

Research and initial survey calls may help you discover some of these problems, but you'll never gather a full picture of any company. Much of this information is privy to the company and not for external dissemination.

LOCATION

You can't help where you live, and if you are physically distanced from those you are soliciting, you are at a disadvantage. Sure, you could move, but if film or videomaking is secondary to a full-time job and life, you can't always pack up and leave. Most distributors are on the West Coast — specifically, the greater Los Angeles area. Budding independent producers who can meet and deliver presentation packages in person have the benefit of being in someone's face. That kind of personal connection is hard to discount in the selling situation. Think about those annoying telemarketers. Their pitch is sure a lot easier to dismiss than someone standing on your doorstep. The same theory applies here. The industry revolves around Southern California. All of the players are located in that area. Being there means you have more access to information, more chances for connections, more time to spend promoting your project to the people that count. Living in Alabama makes it hard to stay in touch, and to be honest, is sometimes a psychological deterrent to being taken seriously. While it sounds harsh, people in the entertainment industry want to deal with people like themselves, not wannabes. That's not elitism; it's just a fact of life that applies to all groups of human beings. Do you think cow wranglers want to spend the day chatting about their trade with a computer programmer from Detroit? It's human nature to gravitate toward people like yourself.

SALESMANSHIP

Not everyone is a salesman. If you don't think you have the ability to, or are uncomfortable with the prospect of calling and soliciting buyers, find someone who can. Writing letters and mailing packages is one thing. Making phone calls to ask someone if he liked and wants to purchase what you made is a whole different venture. You don't have to be slick, but you do have to be able to provide logical and attractive counterarguments when a buyer can't see why your project would benefit his company. Believe me, you are not going to convince anyone to do anything they don't want to. But a good salesman foresees the roadblocks to a sale before they even happen, and steers the conversation in another direction to avoid discussing the downside of a project. When you hear "not interested," you need to be willing to dig for reasons. Simply saying, "Okay, thanks," before hanging up in despair is not going to get you anywhere. You need to be willing to make numerous calls, act politely toward rude people, swallow your pride, and deal with rejection — all the while acting like your independent film or video is the greatest thing your next prospect ever saw. Good salesmen never accept defeat; they look at rejection as a challenge. Can you maintain that frame of mind through this process?

RESOLVE
How long your sales effort lasts is not entirely up to you. Granted, it is your responsibility to make the calls and send the packages, but other issues impacting your performance and ability to hang in there for the long haul can arise. The increasing financial burden of the process may halt your progress. Though not exorbitantly expensive, postage, printing, and long-distance charges add up. If you must halt communications or are forced, for example, to utilize standard mail instead of priority or overnight services, a distributor may move on to something else. Time crunch is another factor. Balancing a day job, family responsibilities, and film and video promotion is tough. Again, a slide in prompt attention to potential buyers' needs due to real-life commitments may find you falling out of favor.

Chapter 15 Summary Points

- Life isn't fair and neither are distribution deals.

- Your timing, lack of sales skills, physical location, and failing resolve can all sink an independent film or video's chances for distribution.

WHAT TO DO?

Just because you don't get picked up by a traditional distributor doesn't mean you and your project will never find an audience, sales, and fame. The opportunity to continue making more films and videos is not lost. In fact, all of these results are attainable — it's just going to take a little more effort on your part.

After exhausting all possibilities of landing a deal with a studio or large independent distributor, it's time to rethink your strategy of approaching the market. No longer can you count on someone else handling all the details of getting your film or video on a shelf or in a consumer's hand. The level of success you achieve now is mostly up to your skill in the art of self-distribution.

HOW DO YOU DO IT?

PART TWO: SELF-DISTRIBUTION

What is self-distribution? It might be easier to state what it isn't:

- Quick
- Easy
- Encouraging
- Impossible

That's right, regardless of how many stories you've been told about the horrors of self-distribution, the most important thing to remember is:

Successful self-distribution of independent films and videos is possible.

Obviously, the other points are of concern. Nobody is knocking on your door to buy rights, and most likely the project hasn't been screened at a festival or received any publicity (or notoriety) of any kind at this point. You'd be dumb to ignore these issues. Self-distribution involves a lot of research, phone calls, letter writing, mailings, waiting, promoting, and — mostly — resilience. Some would say that you are basically starting from scratch, but that is not entirely true.

To your advantage, you already have:

- A solid representation of your project (concept capsule)
- A firm depiction of your audience (audience profile)
- A clear understanding of your goals and limitations
- A working knowledge of researching, surveying, and contacting buyers
- A well-developed presentation package

The process of self-distribution isn't entirely different than what you've been doing to attract a traditional distributor. In fact, if those efforts didn't produce a contract, you can view that necessary experience as very good training for this next step. To self-distribute merely means that you'll be responsible for many of the functions a mainline distributor would have performed for you, including contacting buyers, packaging, promotion, pricing, sales, shipping, and collections.

To be honest, this is a whole business in itself. Functioning as a self-distributor will find you ceasing the creative "making" process for a while, to concentrate instead on finding an audience for your product. This "closure" is vital to you both artistically and financially. It's hard to begin work on another film or video when the last two years of your life were devoted to one that is still sitting on the shelf. Sure, the act of completing a feature movie, documentary, or instructional tape is an accomplishment itself. But most of you purchased this book for a reason. You want and need your project to be bought and viewed by the audience it was intended for. And, unless you are genetically wealthy, it's also not a bad idea to at least breakeven on your investment. The very real concern of making a profit can't be discounted, especially for film and videomakers who are planning on a career in the industry. An unsold project makes it hard to personally justify investing in another one, and virtually guarantees you won't find outside investors. Would you risk your hard-earned money on a film or videomaker who hasn't sold any films?

SEVEN KEY STEPS

Whether you are looking at self-distribution as a way to feed your need for an audience, generate cash flow to pay off past debts, find financing for another project, or just a necessary evil to complete the act of creation, there are certain steps to follow in carrying out the process.

Successful self-distribution involves seven key areas of work for the independent producer:

1. Understanding the various markets available
2. Utilizing the media (free and paid)
3. Locating, qualifying, and contacting appropriate buyers

4. Adapting your presentation package
5. Exploiting other promotional avenues
6. Making the sale and signing contracts
7. Delivering on promises

Each of these is equally as important as the other, and will allow you to present your product in such a way that it can compete with films or videos being released into the system by the studios and large independent distributors.

While an independent self-distributor is definitely at a disadvantage when it comes to financial resources and economies of scale, any worries you might have about limited promotional funds available can be countered by the fact that a good quality product — packaged, promoted, and presented in a professional manner — levels the playing field for most buyers.

THE MARKETS – WHO IS OUT THERE?

Self-distribution will find you promoting and soliciting your product to various markets mainly in the home entertainment industry. This includes video, DVD, cable, and television. Other buyers of self-distributed product can be found in institutional and foreign markets, as well as some ancillary event, festival, and market settings. Theatrical release, a dream for most independent producers, is a tough market to crack without a traditional "connected" distributor. It can be done, and will be discussed, but the emphasis is on software sales opportunities, as these markets provide film and videomakers greater ease of entry and better profits.

Self-distribution affords a lot of entry points; there are no hard and fast rules about when and/or where you jump in. It's now up to you to exploit markets in whatever order you choose. It's still suggested to approach the industry in a top-down manner, starting with theatrical (if so inclined) and moving through home video, cable, and broadcast. But since you'll be doing all the work, you can now make a foreign sale while you are waiting for a domestic video purchase, or rent your project to a local organization while closing on a small video chain deal. Not only *can* you make these kinds of concurrent sales, you *will* want and need to do business in this manner. And the home video/DVD market presents the best prospects for this kind of action.

Home Entertainment Market

Home video, which is the most lucrative part of the home entertainment market, is divided into two segments:

- Rentailers – firms like Blockbuster, Hollywood Video, and local video shops offering product in their stores mainly for rent
- Retailers – firms like Best Buy, Wal-Mart, and Suncoast Motion Picture Company offering videos and DVDs for sale

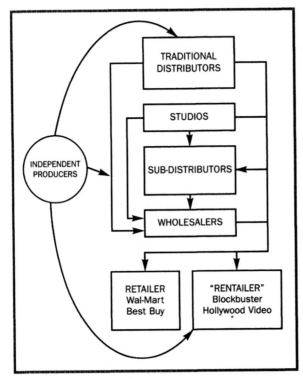

Firms in these two segments get their product from a variety of sources and sell to a variety of customers. The independent producer's role in the process can occur at many stages.

It's important to understand the history of this still fairly new industry, as much of your project's chances — including its ultimate acceptance or rejection — will be based on what happened in the not-so-distant past.

Once a "promised land" for independent producers, the home video market is no longer the easy sell it once was. In the beginning, movies like the $26,000 *Blood Cult* made a killing by selling copies into most every video store on the planet. The home video industry was young and stores needed product. If you could deliver a cohesive feature film, documentary, or instructional video that was adequately packaged at best (I even remember black-and-white video sleeves on the store shelves), you could find a buyer. Actually, you could find a lot of buyers. Unfortunately, it doesn't work this way anymore. What brought about this fall in the low-budget film and video market?

The explosion of the home video market in the mid-to-late 80s, coupled with the simultaneous availability of high-tech equipment at low costs, played a major role in crumbling the opportunity. With these two conditions present for a number of years, anyone could (and did) make a movie or video to sell. The equipment to produce a film was within the reach of many producers, and the hungry market ate up product, regardless of quality. The result was a home video market glutted with inferior product, displeased store owners, dissatisfied customers, and a large portion of distributors growing weary of picking up a project just on the merits of its artwork. Thus, the selling market for independent films and videos has become one filled with doubt and caution.

The good side of this dilemma: Many of the one-shot film and video producers dropped out of the game. Only those with a dogged perseverance to achieve success and the ability to produce acceptable product have survived. While this "flushing" has not created a void of independent product, it has lessened the severity of the problem, and both distributors and their customers (video retail and rental stores and chains) are again in need of more diverse offerings.

Another cause for the suppressed desire of independent material in the home entertainment market can be found with video retailers' obsession with so-called "A" titles. But if you look at the situation closely, an interesting fact emerges: It's not so much the "A" title a video buyer is interested in, but a title that has received a large amount of promotion. The process goes like this: A film or video receives a great deal of publicity and promotion, the public becomes aware of the project and asks for it at the video rental and retail store, the stores want to stock more of those kind of titles. Whether they are truly Hollywood "A" titles or just projects that have been promoted and publicized well (something that an independent film or videomaker is very capable of doing), buyers will continue to demand these products over films and videos with less perceived audience exposure. Put yourself in the buyers' shoes — would you rather have two copies of a well-publicized movie on your shelf, or a couple of independent films that have received little media attention (and are unrecognizable to customers)? It's not a tough choice.

Sources of Product

Technology has forced the home video market to change the way it does business. What was once an industry of fairly simple offerings, sales environments, and neighborhood customers now encompasses multiple media delivery to clients around the world. DVD is quickly eroding the VHS marketplace, sell-through pricing (which is a price point low enough for the consumer to justify purchase over rental) is becoming more popular, chain stores like Hollywood and Blockbuster are eliminating mom-and-pop shops, and the Internet, PPV, and dish-TV services are constantly competing for the home entertainment dollar. Retailers and rentailers coping with this volatile environment, however, still need one thing to stay in business: product. They get that through a variety of methods.

Wholesale Distributors

Though product reaches both rental and retail stores through direct distribution channels, the wholesaler is still the biggest force in this industry. Selling mostly studio and large independent suppliers' products, wholesalers also work with independents, and offer the self-distributor a first point of contact for reaching the home market.

Due to the volatility of the home video business, five companies constitute the majority of the wholesale distribution business:

- Baker and Taylor
- Flash Distributors
- Ingram Entertainment
- VPD
- WaxWorks/VideoWorks

Each has its own set of policies concerning how they do business with an independent supplier. For example, those soliciting Ingram Entertainment will need to include in their presentation package:

- A finished copy or copies of the film or video you wish to submit, which should also include completed packaging and artwork (note that this material will not be returned by Ingram)
- Retail and wholesale prices for that material
- Full terms of business, including dating, freight policies, good and defective return allowances, and origin point of shipping for your goods
- Any pertinent media coverage and/or public relations information on your program
- Any available sales and marketing information on your program (sell sheets, postcards, flyers, posters)

After forwarding this information, Ingram tells potential independent vendors to allow two to four weeks for a response (after which they may still refer you to a subdistributor). This is pretty much standard operating procedure with all of the wholesalers, as well as with Rentrak, the main leasing distributor. You'll want to contact each separately and inquire as to their policies.

Obviously, wholesalers need a variety of materials that you haven't yet developed. That's why you solicited the traditional distributors first. Striking a deal with a studio buyer means you are not responsible for developing things like packaging, freight policies, pricing, press coverage, and marketing materials. Taking care of these details yourself is why the process is called *self*-distribution. Don't worry. All of the items mentioned above are within the grasp of any independent film or video-maker. Creating them is the easy part (I'll explain how in the next few chapters). Finding out where and to whom to send them is what will contribute to a successful self-distribution effort, and should be the focus of your energies at this point.

Wholesale distributors offer product to buyers via catalogs and personal selling. If a wholesaler accepts your film or video, it will be listed, along with terms, in one of these catalogs. As a vendor, you also have the option of purchasing an ad (which is expensive) to increase sales of your product. Since so many video rental and retail shops buy from wholesalers, many independents selling through such an arrangement place their own print ads in trade magazines and use mailers to announce the product's availability through these traditional wholesaling sources.

Once the catalog orders are collected and tabulated, the wholesaler sends a purchase order to the producer, requesting a quantity of product be shipped to one or more specific distribution centers. When working with Rentrak on *Killer Nerd*, 600 copies were sold from the film's first catalog listing. Instead of buying a catalog ad, we used sell sheet inserts for the initial promotion, basically an 8-1/2" x 11" ad that was placed inside of an envelope along with the catalog. It was cheaper than buying a print ad in their publication, and definitely paid for itself. Once we had the purchase order, we were responsible for duplicating, packaging, shrink-wrapping, and shipping the 600 copies of the movie on VHS to two separate distribution centers within 30 days. Payment for the tapes would then be paid to us approximately 90-120 days later. That was after the costs for returns and defective merchandise were deducted.

Though the process of wholesale selling is somewhat confusing, and can be intimidating to the independent producer, a great number of video or DVD copies of your program can be sold in this manner. The best advice I can offer in working with a wholesaler: Follow their submission instructions very precisely. Wholesale distributors are used to dealing with multi-million-dollar supplier companies, and it won't bother them in any way if they have to reject your solicitation because you were unable to follow directions. Don't ruin this opportunity because of laziness or inattention to detail.

When contacting wholesale distributors, check their Web sites first. Most have very precise information regarding submission policies and will appreciate your use of this resource instead of a phone call. If no information can be located, call the firm's headquarters and ask to speak with someone about becoming a vendor/supplier. They'll know what you mean. Wholesalers don't really specialize in genre, so you don't want to survey these folks in detail, just get answers to the following:

■ Does the wholesaler work with independent suppliers?

■ What are its submission policies with regard to program format and length?
■ What materials need to be submitted?
■ Where and to whom should a package be mailed?
■ What is the time period for review?
■ Does the wholesaler work with subdistributors?

As before, be quick, professional, and courteous with anyone handling the call — and be sure to get their name for future reference.

Independent Distributors

In addition to wholesale distributors, many home entertainment stores buy product from independent distributors. These businesses, like Spectrum and Maxim Media Marketing, are often niche marketers, offering either a varied but limited line of films or videos, or a full line of product from the same genre. For example, Spectrum Films markets what could be considered mainly "B" movies. The company does not sell any "A" titles, as its specialty is in edgy and racy independent films featuring sensational or horror-themed subject matter. Full Moon is another good example. Creators of the *Puppet Master* series, the company offers well-packaged sci-fi and horror titles to a loyal audience. Most of the company's films have similar themes, action, settings, and talent. They have, in effect, branded the look of their product so video retailers and rentailers know what to expect from a Full Moon title. The company also sells an erotic thriller line of movies to the home entertainment market.

This independent distributor category encompasses a wide range of businesses — from very small specialty sellers of home repair instructional tapes to the likes of Troma Films, which offers hundreds of films and videos through its huge catalog. With so many companies operating in the market, needs vary widely with regard to what they want from you. Due to their limited size and business volume, many independent distributors, like wholesalers, will want a complete package from the independent producer: finished film or video, packaging, press, and pricing. Others will only be interested in the project, wanting to handle packaging and press on their end.

Undoubtedly, many independent distributors are included in your second- and third-tier solicitation target lists. While initially not as desirable as a studio contract, working with a reputable independent can mean your project will see wide-reaching success. Like the bigger players, these distributors can reach many markets and buyers through their industry connections. Approach these companies much like you would a wholesaler, and since you've already surveyed and are familiar with their submission policies, creating a presentation package containing all of the correct elements should not be a problem. The process is the same: Verify all mailing and contact info, ship all required materials including screener, any packaging, and media, wait two weeks and make the follow-up call. Again, be sure to solicit only those distributors that work with projects like your own.

Subdistributors, those hard-to-define (and find) product sellers also feed the home entertainment market. Working between the product suppliers, traditional distributors, and wholesalers on the one hand, and the video retailers and rental shops on the other, these business- es can move hundreds of tapes in a single sale, often dealing in C.O.D. terms. The advantage to such a set-up is quick cash for the inde- pendent producer, as the bill for the tapes purchased is paid when the tapes or DVDs have shipped. The disadvantage to the situation is that you must have an adequate number of tapes on hand to do business in this manner. When self-distributing *Killer Nerd*, Southwest Tapes, a Dallas, Texas-based subdistributor (no longer in business), started selling the film to small stores across the southwest. The company's first order was for 25 tapes. That was easy enough to get duplicated, pack- aged, shrink-wrapped, and shipped in 48 hours. Over the course of our dealings, the company sometimes ordered upwards of 150 tapes at a time, which was not easy to front-end finance when you consider the cost of duplication, packaging, and shipping. It was a good deal and the firm moved a number of tapes, but we always had to have at least 200 tapes on hand, product that could just as easily not have sold.

Independent Producers

While not a common practice, independent producers sometimes func- tion as direct product suppliers to the home entertainment market. Personally selling tapes and DVDs to video rental and retail outlets is a great way to learn the business, affords you the ability to make numerous contacts, gets your movie or video in the hands of an audience, and can be a profitable way to sell your project.

A sell sheet specifically designed for local producer-to-retailer sales

It's not all "wine and roses" though, and there are many reasons why the process is not only uncommon, but unpopular:

Stores don't like to deal with independent producers – You know this argument. Would you want to buy vegetables from one store, meats from a second, and bread and dairy products from a third? Same theory applies to video buying. The good news: Just like you might be willing to venture to a

small, out-of-the-way store to purchase a specialty item (like good coffee or exotic fruits), so too are video buyers willing to make purchases from small vendors offering something special (known as "niche" products).

It's a lot of work – Researching, surveying, contacting, mailing, and selling to qualified video retail and rental outlets is a time-consuming process. It's not like you can set up a stand and offer your film or video to everyone passing by. You have to *find* your buyers, hunt them down, seek them out. They are not coming to you. You need someone committed to the selling process on at least a part-time basis to make this effort a success.

It can be expensive – When functioning as a true self-distributor, you are responsible for the cost of everything. From dubbing tapes and printing DVD cases to postage, long-distance phone calls, and the value of your time, there is no "parent" company to turn for financial help. Every step of the process must be self- or family/friend-funded.

That's the bad news. The good news is that you *can* successfully self-distribute your film or video in a financially efficient manner and still have a personal life (a small one). By approaching the process in a step-by-step manner, you can control your level of involvement and amount of expense. This method also gives you a chance to "get comfortable" in your new position as film promoter and salesperson, by slowly gaining experience with each task ahead of you.

Broadcast Markets

In addition to the "take-home" markets just discussed, the home entertainment category also includes "in-home" product providers, the largest being broadcasters. For the independent producer, the broadcast market consists of three areas: network, UHF/LPTV, and PBS. None of the national networks (CBS, NBC, ABC, FOX, UPN, WB) buy independent films or videos. They create all of their own programming or get product through deals — that take years to arrange — with studios and other large producers. For now, they are not an option.

Network affiliates, local channels that carry the network programming, do work with independent producers in a limited capacity. They have time slots (usually very late at night and in early morning hours on weekends) open for local programming. This airtime is often filled with affiliate-produced talk and community-service shows, but the opportunity exists for outside independent projects to air.

UHF stations, though sometimes network affiliates, do work with independent producers in acquiring product. Many UHF stations are part of a group of stations (owned by companies like Clear Channel and Infinity Broadcasting); there are very few independently owned ones left. Even so, the same late-night and early morning time slots are usually open (if not filled with "paid programming"), and programming directors at these stations are often willing to work with independent film and videomakers for "one-time-shot" arrangements at airing their work. Series programming (fishing, cooking, or decorating shows) sells best because it offers the director an opportunity to build an audience through a long-term commitment of broadcasting the same show, on the same day, at the same time weekly.

Low power television stations (LPTV) operate in a similar manner to UHF broadcasters, except their signal is only transmitted a short distance (at least with regard to broadcasting). Serving small communities, LPTV stations program a mix of low-cost syndicated shows and locally originated productions. For example, the CAT network of three Cleveland and Akron, Ohio-area LPTV stations buys much of its daytime programming from American One, a 24-hour general entertainment broadcast network owned by USFR Media Group, Inc. in Houston, Texas. The network provides family-oriented programming to over 130 broadcast affiliate stations across the U.S. and over its Internet site at *www.americaone.com*.

In the evenings, the CAT network (located at channels 26, 33, and 35 on the UHF dial) shows independently produced programming that includes a late-night horror movie (with host), automotive sales, and realty and religious talk shows.

Public Broadcasting (PBS) is a tough nut for independent film and video-makers to crack. Created in 1969 to air publicly created and themed programming, PBS channels now compete with the other networks in providing high-quality programming in a non-commercial environment. Headquartered in Alexandria, Virginia, the private, non-profit media enterprise is owned and operated by the nation's 349 public television stations. Looked upon as a trusted community resource, PBS delivers quality programs and education services available in 99% of American homes with televisions, serving nearly 100 million people each week. Though PBS broadcasts mostly independent productions, with its size and reach, the public network has become very selective with regard to program format, genre, and length. Still, the parent station — and especially its local affiliates — offers a potential market for producers willing to work through their maze-like acquisitions processes.

Cable Markets (CATV)
With more than 280 national, regional, and local cable stations and networks across the country, this market provides many opportunities

for the independent producer. Beginning in 1948 as an alternate television service to households where reception of over-the-air TV signals was poor, cable television has since expanded into a multi-billion dollar industry serving more than 70% of U.S. television households. Once simply considered a conveyer of video programming, cable's broadband infrastructure provides an ideal pipeline for delivery of new advanced services, including digital networks, video-on-demand, interactive television, high-speed Internet access, and telephony.

National cable networks such as TBS, Discovery, ESPN, and USA rely on originally produced and acquired programming to fill their schedules. Reaching upwards of 86 million subscribers, the top 25 cable networks work with syndicated-package program suppliers that provide multiple hours of shows and movies aimed at an obviously wide audience. Though many of these top cable networks support a specialized genre (sports, women's issues, music, weather, news), programming is still very mainstream and aimed at a somewhat broadly defined audience base. In their infancy, national cable networks were once a target for independent filmmakers whose product was used to fill late-night and off-hours time slots. The USA network is a good example. The ultra-low budget *A Polish Vampire in Brooklyn*, which reportedly only cost $15,000, was a staple of that network's late-night programming. This is no longer the case, as paid programming and dated Hollywood fare occupies many of the movie-based network's hours.

Because of advancements in cable technology, *specialized networks* have experienced amazing growth to fill the 500+ channel capacity of most providers. Offerings vary widely and include networks such as:

- TechTV, the only cable television channel covering technology information, news, and entertainment from a consumer, industry, and market perspective 24 hours a day.
- WISDOM Television, which features unique programs and original series focusing on the areas of personal growth, social and environmental consciousness, and healthier living.
- Native American Nations Program Network, with programming that includes entertainment, news, sports, business, specials, and talk shows.

■ Hobby Craft Interactive (HCI), with instructional, educational, and informative "how-to" programming aimed at various crafts and hobbies ranging from quilting, home sewing, knitting, and crocheting to painting, woodworking, ceramics, and stained glass to model railroads, car restoration, antiques, and collectibles. HCI also features fine arts programming with visits to museums and studio tours with working artists.

Obviously, with this breadth of subject matter, a number of opportunities exist for independent film and video producers with product to sell.

Regional and local cable networks offer programming specific to a limited geographic area. Many of the operators are state-based, such as:

■ Pennsylvania Cable Network (PCN) – offering live and same-day coverage of the Pennsylvania Senate, House, and other governmental activities. PCN also televises significant state events (such as high school sports championships), tours of museums and manufacturing facilities in the state, and distributes educational programming.

■ Ohio News Network (ONN) – a comprehensive 24–hour cable news channel featuring local news, weather, and sports for the people of Ohio. ONN offers a lineup of innovative and interactive programming, including *Ohio's Talking, Ohio Sports Insider, High School Sport Site, On the Square, Ohio's Most Wanted, Ohio's Missing Kids,* and *ONN Morning Journal,* and the only live, continuous coverage of the state legislature.

Some of the hundreds of specialty national and regional cable channels and networks available for programming solicitation:

ActionMax
All News Channel
AMC (American Movie Classics)
AMC's American Pop!
American Legal Network
America's Store (National Cable)
Animal Planet
Anthropology Programming and Entertainment
Anti-Aging Network, Inc.
Applause Networks

Arabic Channel
ART (Arab Radio & Television)
Auto Channel
BBC America
Beauty Channel
BET (Black Entertainment Television)
Biography Channel
Black Belt TV/The Martial Arts Network
Black STARZ!!
Black Women's TV
Bloomberg Television
Boating Channel
Bonjour USA
Booknet
Box Music Network
Bravo, The Film and Arts Network
Celtic Vision
Collectors Channel
College Entertainment Network
Courtroom Television Network (Court TV)
Crime Channel
Discovery Health Channel
Discovery Home & Leisure
Discovery Kids Channel
Discovery Science Channel
Do-It-Yourself Network
Dream Network
E! Entertainment Television
Ecology Communications
Ecumenical Television Channel
Fashion Network
Filipino Channel
Food Network
Free Speech TV
FX (Fox Basic Cable)
Video Gaming Network
Game Show Network
Hip Hop Network
History Channel
Hobby Craft Interactive
Home and Garden Television (HGTV)
Home Shopping Network (HSN)
Independent Film Channel
Michigan Government Television (MGTV)
Military Network
My Pet TV
NASA Television
National Geographic Channel
National Greek Television (NGTV)
National Jewish Television
Native American Nations Program Network
Outdoor Channel
Outdoor Life Network
OVATION - The Arts Network
Oxygen Media

PAX TV
Puppy Channel
QVC
RadioTV Network
Rarities-Exchange
Real Estate Network
Scandinavian Channel
SCI FI Channel
Seminar TV
Senior Citizens Television Network
Shop at Home
Soapnet National Cable
Speedvision Network
Spice 1 National Cable
Sundance Channel
Sunshine Network
TBN - Trinity Broadcasting Network
TBS Superstation
Tech TV
Theatre Channel
ThrillerMax
TNN
We: Women's Entertainment
Weather Channel
World Cinema
Worship Network
Yankee Entertainment and Sports Network
(YES)
Yesterday USA
Youth Sports Broadcasting Channel

In addition to the state-run channels, cable providers (known as MSOs or multiple system operators) like Time-Warner, Cox, Comcast, and Adelphia must offer some local channel availability through each origination point. Chances are you've flipped through some of this "local cable" on your own television; it usually deals with city council meetings and bulletin-board type announcement pages. That same time is open to other programming as well.

Leased access was created by the Federal Communications Commission in an attempt to rectify the complaint that the carrier of goods (in this case the MSO), providing service to other producers, cannot also own goods, or at least all the goods. The carrier must be free of conflict if it is to provide carriage to all. This common carrier deregulation concept, originating with the railroad system, carried over to the telegraph and the telephone industries, and in the mid-90s, to the cable industry.

It's a tough market to navigate, as cable operators aren't always willing to work with leased access producers, but the law says they have to comply. From infomercials to pay-per-view events to talk shows and movie programming, independent producers working with leased access can reach local, regional, and national markets by literally leasing the airtime. Rates vary depending on the cable system, number of subscribers, and time of programming, but it's very affordable. A half-hour block of time can be leased from Cincinnati's Time Warner cable system, which serves more than 350,000 subscribers, for $150 in the 7pm – midnight time slot. You can buy as little as a half-hour block or lease a station 24 hours-a-day. Leased access has been a hotly debated topic in the cable industry for years and is a little-known avenue for producers looking for cable markets.

Pay-Per-View/Video-On-Demand

For consumers who want to watch a new movie without the hassle of returning a tape or DVD, or even leaving their home, pay-per-view (PPV) was created. Available through all traditional and digital cable systems and satellite television, pay-per-view is usually available on at least six and sometimes up to 25 or more channels in a system. Consisting of mostly new movies, sporting events, wrestling, erotic specials, and music concerts, PPV programming is available 24 hours-a-day with rotating content from such services as:

- Encore
- inDEMAND
- Playboy TV
- Spice
- Starz

While these companies have partnerships with major Hollywood studios and get most of their content from those relationships, many are exploring more niche-oriented fare and looking to independent suppliers to supply programming needs. This is a technologically dependent market that changes quickly with regard to audiences and delivery methods. As services expand and become more targeted to specific customer wants, so too will programming needs expand.

Video-on-demand (VOD) is identical to the in-room movie service found in many hotels, only now it is available in the consumer's home. Viewers can watch a variety of programming, including new movies, favorite television programs, events, concerts, and sports exactly when they want. New technology even allows the viewer to pause, rewind, restart, and fast-forward the programming at any time. Video-on-demand is also very dependent on technology, and many partnerships with traditional cable or satellite delivery networks attest to the growing future of this film and video distribution service. And, like pay-per-view, niche markets are an increasing focus of many VOD operators, as the service allows for economical delivery of specialty programming to a broad audience.

Institutional Markets

Though the home entertainment markets of video, DVD, broadcast and cable television, and new pay-per-view and video-on-demand services often offer independent self-distributors the greatest opportunities for both exposure and profit, many producers find equal (and sometimes greater) success in the institutional marketplace. It all depends on the product.

Feature-length fictional films, some documentaries, and many instructional videos are best suited for the retailer/rentailer distribution scenario of the home entertainment set-up. These are the kinds of things people like to watch in their homes, at leisure. The institutional market, however, attracts a different audience, one more interested in learning, teaching, and examining subject matter or issues relevant to the group's beliefs or reason for being. Thus video productions concerning very specific topics or aimed at specialized interests, hobbies, culture, geography, or individuals, are perfect for distribution within the institutional setting. For purposes of this book, we'll look at four distinct segments that comprise the institutional setting:

- Schools/Universities
- Libraries
- Institutions
- Organizations/Corporations

Each of these groupings host unique needs with regard to format, length, subject, and audience for a production.

Schools/Universities – Anyone who has spent anytime in a classroom — be it elementary school or college — knows that video is playing a bigger and bigger role in the teaching process. Depending on the class and grade level, instructional, educational, and informational programming of all kinds is used to enhance the learning experience. This potentially huge market (considering all of the public and private educational institutions in the United States) can be further defined by region (local, state, national), audience served (pre, elementary, middle, and high school; undergraduate and graduate colleges), and subject matter (academics, social skills, athletics, special interests). A system of specialized educational distributors is in place that serves much of the market (i.e., *www.teachersvideos.com*). These businesses acquire product through a network of independent suppliers, and sell to schools and universities with a combination of catalog, online, and personal calling approaches. Mirroring the home entertainment market set-up, a producer's chance for both wide exposure (to this specific audience) and financial success is increased when a project is acquired by an educational media distributor.

Libraries – Sometimes referred to as another home video market (since patrons can borrow videos along with books), libraries present independent film and videomakers an accessible and supportive outlet for their work. Like their paper-based offerings, libraries get videos and DVDs from specialty distributors serving that market. Depending on the buyer's needs, direction, budget, and — quite honestly — taste, programming is selected and purchased on a fairly regular (weekly or monthly) basis. Approved distributors make up the bulk of video and DVD sales to the market, though the independent trade is very brisk when it's available and applicable. For example, *Pig* was sold direct to library systems without the services of a traditional distributor. The movie offered local interest both for its shooting locations and cast and crew. The film also appealed to libraries in fulfilling their dedication to the promotion of local arts. Most regional libraries house extensive collections of books both written about the area and by authors from the area. The same theory holds true for independent films and videos. Instructional and educational videos are always a natural for this market, too.

Institutions – In addition to schools, universities, and libraries, a number of other institutions exist that are viable candidates for an independent self-distribution effort. The highly populated prison system is just one example; there are distributors that specialize in serving this "captive" market. Entertainment, educational, and instructional films and videos are brought into facilities for a rental fee. Distributors source most of their product from Hollywood studios, as blockbusters prove to be the most popular with audiences. Independent fare, both features and instructional product, have been gaining an increased presence though, especially in minimum-security environments where inmates have more freedom and access to entertainment.

After-school daycare and senior centers are two other institutional markets aggressively utilizing video to serve their clients. Though some educational distributors and service providers exist that serve these categories, a truly organized distribution service is not really in place, since these institutions are often hard to track. Other institutional markets offering sales opportunities to independent producers include:

- Churches
- Mental health centers

- Hospitals
- Community recreation centers

Organizations and corporations – Too numerous to count, organizations are always looking for entertainment, enrichment, or education to share with their members. Independent videos dealing with subject matter specific to the group's interests or goals are usually well-received and used in a number of ways. The material can be bought for a public showing, listed for sale in a group newsletter, or offered at a meeting. Some specialty independent distributors exist that serve extremely defined segments (hunting videos to sportsmen's groups, war videos to veteran's associations, nature tapes to environmental clubs), but like the institutional market, there is no official way of "doing business." Some of the many organizations to consider as potential buyers for your films or videos:

- Locally based groups (Rotary, Kiwanis, Lions, Jaycees)
- Veterans associations (Veterans of Foreign Wars and American Legion)
- Educational associations (parent-teacher associations, literary guilds)
- Social groups (women's auxiliaries)
- Culturally based clubs (native and African American groups)
- Specialty clubs or groups (gardening, civil war enthusiasts)
- Issue-based/activist societies (right-to lifers, People for the Ethical Treatment of Animals)
- Professional fraternal organizations (police, firefighters)

The corporate environment is another market utilizing video in its day-to-day operations. Buying and renting videos mostly for training purposes, companies are always looking for new and improved ways to work, teach employees to interact, and orient new hires to the workplace; videotape/DVD provides an effective and affordable option. Instructional presentations on public speaking, phone skills, computer skills, interviewing, and motivation are perpetual bestsellers. The market, served through a variety of suppliers including specialty distributors and independent sellers, is also well fed through mail and online ordering.

Foreign Markets

A very lucrative market for Hollywood product (some sources say upwards of 40% of a film's total revenues come from the foreign market), non-domestic sales also present a good opportunity for the independent film and videomaker.

First, know that when people are speaking about selling foreign rights to a project, they usually means anything outside of North America. For purposes of this book, however, and considering that you'll be distributing the film or video yourself, foreign means anything outside of the United States (as it should). The reason: You just can't box up some tapes and ship them to Ontario. While you may not need the services of a broker to make a Canadian deal, the process is much different than making a sale to a store in Wisconsin.

Navigating the international market without a distributor or foreign sales agent is tough and often impractical. The selling of foreign rights is such a broad-based and specialized field that most small and mid-sized distributors hire a sub-agent specialist to do the job for them, and the same applies to you. It's truly impossible to know what kind of product is right for the various markets. Questions of salability include:

- What genres (horror, western, comedy, instructional) are popular in the country?
- What kinds of ratings systems are used?
- What formats (feature-length, shorts, documentaries) are appropriate?
- What media (video, film, digital, streaming) is acceptable?
- What markets (theatre, home video rental, retail, television, cable, Internet) are available?

Foreign markets exist for nearly every kind of product (*Killer Nerd* was sold to Malaysia and Mexico in early foreign deals), but the independent producer functioning as a self-distributor has no access to the kind of market information needed to make the right decisions in contacting buyers and soliciting sales. Contact a foreign sales agent (much as you would solicit a traditional distributor) to help gauge the level of foreign appeal for your film or video.

Other Markets

Self-distribution means you are always on the lookout for new avenues to reach an audience and turn a buck. You cannot merely depend on video, cable, and foreign sales to put you "in the black," as quite honestly, these markets could end up producing disappointing results, or even worse, returns. If and when those opportunities vanish (sometimes this happens before they even start), you need more revenue streams — more ways to get your film or video "played-and-paid." The self-distributor has to be willing to explore every potential market that exists. Some of the more popular untraditional selling experiences follow:

Theatrical rental – As mentioned earlier, this means of distribution is often referred to as four-walling because the independent film or video-maker literally rents the four walls of a movie theater directly from the theater owner. You are responsible for providing your own print of the film (or video) to project, placing ads in newspapers and posters around town, and sitting in the box office collecting ticket sales money for yourself. You cut out the distributor's middleman expenses, but add a ton of work. This process can be carried out successfully by feature-length film or documentary producers (the *Billy Jack* series in the 1970s is the most famous example), though it requires a lot of legwork. Not all theatres are open to this arrangement; usually it's something that only independently owned movie houses will accommodate.

Local sales – In addition to selling your tape or video to the retailers and rentailers in your town, the community at large offers another market. Though many potential viewers interested in the film or video will check out a copy in the town's video store or gift shop, another segment of the local population may be just as interested in the works of a "homegrown" artist, and be willing to shell out some cash to see your work. Depending on the size of your community and subject matter of the production, some local "markets" that exist include:

- City and county fairs (if product is general interest or dealing with community)
- Local four-walling shows (rent a library or community meeting room for screening)
- Regional art/craft, sport, or home improvement expos (perfect for instructional tapes)

Anything goes here. If a sales opportunity in the local arena presents itself, take advantage of the chance to promote yourself and your product.

Non-profit/Charities – Sales to such groups is very subject-dependent and not applicable to all independent films and videos. You are essentially looking for those non-profit and/or charitable organizations with members expressing an interest in the topic of your film or video. This can be thought of as an add-on market to the organization/corporation market explored above. Like PBS selling copies of its programming immediately following a broadcast, you may find product sales to individuals immediately following a public or group showing. Investigate the same issue-specific groups, charities, clubs, and organizations outlined earlier.

Online – With the unyielding growth of the Internet, the online market constitutes many audiences and numerous sales opportunities previously unavailable to the independent film and videomaker. The low-cost, do-it-yourself aspect of the technology, where one person can literally create and run an online film or video "store," makes it even more beneficial to those projects being self-distributed. Worldwide buyers of all kinds, from individual niche-interest consumers to online media distributors, can be located, surveyed, contacted, solicited, and sold via an Internet connection. This phenomenal means of doing business is explored in depth in Chapter 27.

Event Sales

A final group of sales opportunities lies in those situations where you or your project is the reason for a gathering. Events, whether industry-sanctioned or self-created, present the self-distributor some of the best routes to market his project. Depending on the occasion, audiences can include film and video professionals, distributors, video buyers, media representatives, and fans. The outcomes of an event are just as varied; product sales, distribution contract offers, foreign deals, agent contacts, media exposure, and just plain fun are all possible.

Sales markets – Though many film and video markets exist, those involved with self-distribution rarely have the funding required to attend these high-buck affairs. Just setting up a table at the American Film

Market, for example, can run upwards of $10,000! There are, however, several events taking place around the United States that do offer independent promoters the chance to mingle with and make sales to the various film and video buyers:

Independent Feature Film Market (IFFM) – Held at the Angelika Film Center in New York City, the IFFM is perhaps the largest single sales gathering of independent filmmakers in the world. The annual event showcases more than 300 narrative and documentary features, scripts, works-in-progress, and shorts for the film industry's domestic and international buyers, distributors, development executives, and film festival programmers. Held over six days, the IFFM costs $400 to enter, which gets you a screening of your feature, publication of the film's synopsis, and contact information in the event's catalog and Web site, two VHS cassettes of your program in the event library (for interested distributors who can't make the screening), and access to all of the market events including other screenings, seminars, and social events. With its total emphasis on the independent producer, the market can deliver — in 2000, 21 of the 52 fictional features screened at the IFFM were acquired for some sort of theatrical, television, home video, or international distribution.

Regardless of what the literature says, it's usually best to attend markets with finished product. Mike Mongillo traveled to and screened a three-minute trailer of The Wind at IFFM's Works-in-Progress, an expense he believes would have been better spent on postproduction. "Although those in attendance were impressed and many meetings ensued with potential 'finishing funds coordinators,' not one dime was raised as a result of those contacts. It really is a market of buyers looking for movies ready for market."

Created by the Independent Feature Project (IFP), an independent film-making organization with offices in five cities across the country, the IFFM is really a must for any producer serious about making the contacts needed to find distribution for his film (as is joining the IFP). Every self-distributor needs to find room in his schedule and budget for the week-long sales conference. Exhaustive details about the event are posted at *www.ifp.org*. There you'll find registration information, publicity tips, market strategies, how to make deals, and what to do when the market is over.

Video Software Dealers Association (VSDA) – This group's annual convention in Las Vegas is for anyone doing business in the home video industry. It's a great showcase for debuting new product to individual store and chain buyers, the entertainment media, and traditional studio and independent distributors of all sizes. Many independent producers use the occasion to offer their latest film or video to a convention center full of buyers for a three-day period. Both affordable and friendly, the VSDA recently began an "independents showcase," allowing independent filmmakers to make special distribution deals with large independent video chains across the country. At this event-within-an-event, filmmakers screen their films to buyers much like a traditional film market set-up. Videotapes and DVDs are sold right on the show floor throughout the convention in an almost flea market-like environment. Regional chapters of the national organization hold similar, smaller events that encourage the same kind of independent spirit. The association's Web site, *www.vsda.org*, offers comprehensive information on all aspects of the show including registration, attendee lists, and application and contact information for the independent showcase. You can also contact the VSDA direct at (800) 955-8732.

East Coast Video Show (ECVS) – This is much like the VSDA affair, only held on the other side of the country. There is a definite difference in the audiences (it's that New York-L.A. thing), though again you'll find real support for both your project and attitude. This show is newer than the official VSDA convention, but has grown in both exhibitor size and attendance since its inception. Touted as the country's largest regional trade event for buyers and sellers of prerecorded video products and service, the ECVS is cheaper than the VSDA and features an attendee list that includes representatives from TLA Entertainment Group, Video

Hut, and Video USA Entertainment — all respectably sized video store chains that support independent features.

Other events – Premieres, publicity stunts, public appearances, and film festivals are other events that really constitute a market opportunity in which to sell your film. Each of these techniques are good vehicles for a self-distributor to use in reaching audiences and the media and drumming up sales. Look at Chapter 21 for other promotional approaches and detailed information on event tactics.

Once you understand the markets available to you as a self-distributor, you're going to need some help attracting interest from those same markets for your project. To do that, you'll need to exploit the media. We'll tackle that subject next.

Chapter 17 Summary Points

■ Self-distribution — while not quick, easy, or encouraging — is possible and is often prompted by the very real concern of making a profit on the time and the money invested in making a film or video.

■ The home video market of rentailers and retailers (those firms renting and selling tapes and DVDs) offers the largest sales opportunities to the self-distributor.

■ Though home video buyers are once again looking for a diverse array of product, the market is not the heyday it once was for independent producers.

■ Wholesale distributors maintain a commanding presence in supplying product to the home video market, and many are willing to work with independent suppliers offering quality product.

■ Independent distributors also heavily supply the home video market, though they often tend to specialize in product genre.

■ Independent producers can sell directly to video rental and retail stores, though not as easily as an established distributor.

■ In addition to the "take-home" entertainment buyers, opportunities exist for independent program sales to the hundreds of broadcast, pay-per-view, and cable television outlets.

■ Libraries, schools, and other institutions offer a lucrative outlet for independent producers selling specialty instructional or educational productions.

■ Attempting to broker a film or video sale to the international market without an agent is tough and impractical.

■ Once all traditional markets have been exploited, independent producers can look to theatres, charities, online opportunities, and event sales as further sources of exposure and income for their project.

UTILIZING THE MEDIA

We live in a media-obsessed society. Magazines, radio, television, and the Internet bombard our minds second-by-second with messages of products we can't live without and news that will change our lives. It's hard to believe that in our not-so-distant past, it took upwards of a week for major news to reach the general public. A good example: World War II updates were played as newsreels before a movie. The footage in these productions was often a week or two old by the time audiences viewed it. And people living away from urban areas might go for months without a newspaper story or radio program to inform them what was happening in the world.

Quite obviously, that's all radically changed. Handheld palm computers are now capable of connecting to the Internet, accessing instantaneous news updates supplied by the multiple information providers across the globe. Hundreds of broadcast and cable sports, entertainment, and specialty news programs are airing 24 hours-a-day. And the explosion of niche publications means someone interested in lawn tractor racing probably can find a magazine dedicated to that subject.

The airwaves, news pages, and electronic highways, already over-crowded with information — both good and bad — are about to get worse. And that's because these communication carriers will soon be transporting your messages to the masses.

Like the biggest Hollywood epic, most successful instructional tape series, or history-shattering documentary, the media also carry a high degree of responsibility for a self-distributor's ultimate success or failure. Whether it's paid or free, utilizing the media in as many appropriate forms as possible is not an option — it is a necessity — and you must look to all forms of the media as a "partner" of sorts in your promotional efforts.

WHEN TO CONTACT?
Let's get the tough question out of the way first: At what point do you contact the media? Lots of theories exist as to when this should happen.

Some say as soon as you start your film or video, call up all the local papers, any regional entertainment tabloids, network affiliates in your broadcast area, national entertainment media like *Entertainment Magazine, Ain't It Cool News, Entertainment Tonight, Premiere...* the list goes on and on and on. Others will tell you to hold off on any press contact until the project is complete. Their rationale is if the media covers your independent film during production, they may decide it's old news and not revisit the story later. The argument here is that press is more important when a film is finished and the producers are looking for an audience and buyers. A two-year old newspaper story isn't going to drum up much business when the film is finally ready. But then again, how do you build a press kit for the presentation package if you haven't done any media until you're finished?

The answer to this dilemma often depends on the specifics of the project and the many details involved in its production. A mainstream Hollywood feature boasts the kinds of stars, drama, and budgets that keep the public interested in its development over the course of a year, so there's constant media coverage from start to finish. On the flipside, an instructional video made for $500 should probably hold off on any press until the product is ready for market.

Remember that in traditional distribution arrangements, the filmmaker is not responsible for creating and coordinating the media campaign. But we're talking self-distribution at this point. It is now up to you to generate any kind of "buzz" (or at least "hum") in the print, broadcast, and Web-based media for your film or video. It's a big world out there — how do you know the right time to launch your epic through these promotional avenues? While that's not an easy question to answer, here are some basic guidelines with regard to the various media available to get your activities headed in the right direction:

Local Press Announcements
Announcements in the newspaper are fine throughout production of a shoot. Local media usually rally behind a filmmaker, especially if you are not producing in a major metropolitan area. Small towns don't often host film crews, so continuing coverage of the project won't be a problem. These outlets are especially important during complicated location

shoots, when you may need to smooth concerns and possible tensions of neighbors or businesses near the action. While a small-town newspaper feature on your film isn't as impressive as a *New York Times* review, press is press; it helps build credibility for your film or video as well as providing fodder for the press clippings section of your presentation package.

The same can be said for *local broadcast television and cable stations*. Best served by "on set" visits, as these present more interesting visual opportunities for the camera than talking-head interview programs, any local broadcast and cable media can be solicited at the very outset of production. Like their partners in print, they'll be willing to follow your story throughout its entirety.

A variation: *local radio interview programs*. Always on the lookout for hometown celebrities, radio stations can be approached before a big shoot (to drum up support or extras), during any premiere-type release event, and concurrent with local sales efforts. It makes for good radio to talk of filmmaking war stories and the inevitable goof-ups that occur during the course of any project.

Entertainment Trade Publications
These can be solicited at the very outset of production and at completion. When your film or video enters the production stage, you'll want to "announce" this fact to the industry trades, as each has a special "in development" or "in production" column that is read regularly by buyers and distributors of all kinds. Specifically contact:

- *The Hollywood Reporter* – (323) 525-2000
- *Variety* – (323) 857-6600

Call each and explain that you'd like to get a new project listing. A form will be faxed/e-mailed to you asking for title, format, length, director, producer, cast, production company, and other associated details. Fill it in exactly as it asks and your project will appear alongside big Hollywood productions — all for free.

These same publications can be contacted anytime you have legitimate news on your project, such as its completion, acquisition by a distributor, sales milestone, celebrity cast or crew connection, screening announcement, or other similar information. The trades are dedicated to informing the film and video industries of what's happening; stories concerning your project can appear at anytime and can only help your promotion campaign.

Film Finders

Also at the onset of production, or as early in the process as you are able, you'll want to list your project with *Film Finders*. Though not technically "media," it can be categorized as such by default since the organization disseminates information to a large group of people. Again at no cost, information about your film will be available to domestic and international film buyers, acquisitions executives, festivals, and markets. You can download a submission form and extensive details about the service at *www.filmfinders.com.*

Home Entertainment-Specific Trade Publications

Publications such as *Video Store Magazine* and *Video Business* should only be approached when your film or video project is complete. They do not run production stories, focusing only on what's new and available for sale. This is a good outlet for "free" advertising to many important industry buyers. Reviews of videos and DVDs fill the pages, along with ads for the same. The magazines (and their Web-based versions) are an ideal environment for announcing a product ready for the shelves.

Film or Videomaking-Specific Publications

Magazines like *MovieMaker* and *Independent Filmmaker* cater to folks like yourself and can be solicited for stories at anytime throughout the production. Though many readers are doing the same thing you are — producing and promoting a film or video project — a portion of these publications' readership is comprised of buyers, distributors, agents, and other industry "players." Interesting production stories, interviews with the "next big things," and features on marketing techniques are all possibilities that can be pitched to the editors regardless if your film or video is in preproduction or rolling off the Avid.

Other Film and Videomaking-Specific Media

The same rules apply to other film and videomaking-specific media such as Web sites, cable programming, and the one or two radio programs airing such content. You'll want to contact these outlets as soon as you have something considered news- or feature-worthy. *SplitScreen*, the Independent Film Channel program, did a story on the pair of filmmakers responsible for *The Blair Witch Project* long before it was a national phenomenon. In fact, the story led to the filmmakers' union with a well-known producer rep who helped shepherd the project to the big screen.

General Entertainment Publications

Publications such as *Premiere, Movieline,* and *Entertainment Weekly* are mostly filled with star profiles and blockbuster stories to ensure appeal to the widest possible audience. That does not mean they are averse to covering low-budget independent films. Wait until your movie is very near, if not complete, before contacting these publications. This is the kind of attention you'll want when your project is ready for sale. Do likewise with corresponding sister broadcast and Web-based media.

Industry-Specific Publications

You may or may not know it, but every industry supports its own product with industry-specific publications. Whether they're newsstand consumer magazines or trade periodicals, regular monthly issues are produced to inform, educate, and entertain a devoted and interested audience. From *Tire Review* and *The Welding Journal* to *Soybean Digest* and *Onion World*, every market is covered. More suited for those self-distributing instructional or informative videos and documentaries, these media should be contacted twice: initially to announce production and then when the project is complete. Since these magazines only carry news that is relevant to their industry, they'll really only be interested in the fact that you have made a video that addresses the concerns of their readership. An instructional video on gun cleaning is not only of interest to the editors of mainstream gun books like *Guns & Ammo*, but also to those publishing *American Firearms Industry Magazine*, which supplies news and information to gun shop owners around the world. Again, treat all corresponding industry-specific media in the same manner.

The bottom line here: Approach small, local media as soon as possible to begin accumulating media coverage and hold off on any major press outlets until the film or video is finished. You'll want that coverage only when you have something to sell. There are exceptions — say if you happen to talk Tom Cruise into a cameo for your *How To Sell Used Cars* instructional tape, that'd be big news. But for the most part, don't blow your opportunity to score national attention unless the project is complete. You'll need all the exposure you can get when the self-distribution process begins.

As you begin the process of working with the various newspapers, magazines, television shows, and Web sites out there, it's vital to remember you really only get one shot at most media outlets. A "no" really means "no." Unless you like being on the receiving end of an annoyed discourse on your lack of professionalism and the deficient merits of your so-called media story, believe me and don't test this theory. When a reporter or press outlet refuses or ignores your calls and submissions — don't resend, recall, or retry to get the same information in the newspaper, magazine, Web site, or broadcast outlet through another individual.

Before you start contacting anyone, you need to know who and what is out there that wants and needs your kind of news.

RESEARCHING AND SURVEYING THE MEDIA

The process of finding the proper media to solicit with news on your self-distributed project is not much different than when you went looking for a distributor. The basic steps:

- Collect information on every possible source of media that may cover your project.
- Contact each and survey as to its wants and requirements.
- Sort according to medium (print, broadcast, and Web) and desirability.
- Create a timeline or calendar indicating lead times and deadlines for submission.

It's not hard to find media these days. In the past, self-promoters had to trudge to the public library to examine the *Guide to Periodicals* for a listing of magazines sorted by subject matter. And that was just the start. One still needed to find information on televisions shows, newspapers, and radio programming. This was a long and time-consuming job. You can now thank the Internet for streamlining this step. One of the quickest and easiest means of collecting media sources for your promotional campaign is through online searches. Try keywords like: film/videomaking magazines, entertainment periodicals, film/video-making Web sites, movie magazines, entertainment programming, and cable television networks for access to hundreds of appropriate media. Conduct searches according to your specific genre, as well. For example, producers of instructional fitness tapes will want to surf fitness and health sites on the Web. Many will have links to periodicals supporting that area of interest and the site itself may end up functioning as a media outlet for your news.

A visit to the neighborhood newsstand (or book superstore like Borders or Barnes and Noble) is another good source of information. Check out their racks of entertainment, film, video, and subject-specific magazines. Like the note-taking exercise in the video stores, you'll want to copy contact information (editor, address, phone, fax, e-mail, and submission policies) from any appropriate magazines. This is usually found on the first couple of pages of a publication, near the contents. Again, speak with a manager or store associate before pulling out the pen and paper.

Unfortunately, you can't avoid the library altogether. To gather current information on newspapers and broadcast media, ask to see the *Bacon's Guides*, the most comprehensive listing available to all forms of media. The company publishes four main volumes: a newspaper/magazine directory, a radio/TV/cable directory, a media calendar directory, and an Internet media directory. Eight other industry-specific guides are also available. The 12 books list more than 70,000 media outlets throughout North America and Europe, with information that includes:

- Nearly 500,000 editorial contacts
- Over 100,000 editor-pitching profiles, including preferred contact methods, best days and times, beat specialties, and deadlines

- The largest collection of editor-direct e-mail addresses, fax numbers, and phone numbers available
- Mailing addresses and Web sites

You could purchase these volumes yourself, but they're very costly — ranging in price from $275-$375 each — and the information changes all the time. Libraries subscribe to *Bacon's* and thus have the most current data available. You'll want to spend some serious time pouring through these books, as you'll be surprised at the number of media available — applicable to your project's news — that you didn't even know existed. Don't forget to take along a pocketful of dimes for the copier, too.

Finally, let your fingers do the walking through the yellow pages. Newspapers, regional magazines, television stations, and cable outlets will all have listings there, providing a quick start to contacting local media.

MEDIA CONTACT SURVEY

Working from your list of media outlets, call or e-mail whoever is designated the contact person for editorial submissions. You'll want to ask about:

- Preferred method of submission (e-mail, fax, snail mail)
- Callback policy (do they want you to call after sending info?)
- Lead-time (how long do they need information before use?)
- Type of information preferred (news release, video press kit, b-roll footage, photos, audio tape)
- Correct spelling and pronunciation of writers'/broadcasters' names
- Pertinent e-mail, fax, and phone numbers

Even though much of this information is provided in detail via the Web and media directories, personal contact is valuable to establish a connection. Have your questions ready and phrased for short responses when calling. The same goes for any electronic correspondence. Editors, program producers, and Web developers are busy people who don't have a lot of time to shoot the breeze with a neophyte public relations wannabe. If submissions guidelines are clearly posted, still make the call. Just introduce yourself, express interest in working with that specific media outlet, and ask if any policies or contact information is different from what is posted on their site. This less-than-two-minute conversation is a good door opener, and gets your name in front of someone who is probably at least partially responsible for that outlet's content.

Once you've compiled relevant information on the various media available for publicizing your project, sort according to type and level of desire. Let's say you've created a video on the renovation of antique cameras, and found its very targeted potential audience is part of a group served by a slew of media resources, including magazines, cable programming, and a couple of dozen Web sites. Sorting your data, you find six newsstand magazines serve the entire general photography market. While any press on your video would help your self-distribution endeavor, you find through survey calls and research that only two of the magazines cover renovation and repair issues in any detail. Those publications are obviously the starting point for the majority of your publicity efforts.

TIMELINE

Media research produces a wealth of information. To keep it all straight, and make sure you get the most coverage possible, you'll want to create a media timeline, a guide allowing you to see the big picture all at once, instead of concentrating on the publicity efforts in a case-by-case basis. The biggest advantage of such a timeline: tracking lead times.

Lead-time is the amount of time between a media outlet's release date (when a magazine is mailed to subscribers or placed on the newsstand, when a television show airs, when information on a Web site goes live) and when that outlet needs any creative content to meet that date. For example, *Premiere* magazine may need your media kit in August to prepare for their October issue. Nationally sold magazines will want at least three months lead for submissions, newspaper feature-article writers appreciate a week, while news reporters can work overnight; Web sites can also function with a quick turnaround (no printing or production); and broadcast television, cable programs, newsletters, and trade magazines vary according to their schedules (weekly, monthly, bi-monthly, quarterly). Your willingness (and ability) to get information to the media according to their schedules can mean the difference between getting press or getting ignored. And, by tracking and adhering to these lead-time policies, you'll get press when you want. It does no good to have your late October-released and Halloween-season targeted independent feature, *Horror of Hocking Hill,* reviewed in the March issue of *Premiere* magazine. While any publicity is good, properly timed publicity is great. This media timeline should eventually be integrated into a master schedule that tracks all of the publicity, advertising, promotional, and even sales efforts throughout the life of your project.

GETTING FREE MEDIA
Free media is usually known as publicity. Much of what you read in magazines, hear on the radio, see on TV, and surf on the Net is a product of publicity, or — more specifically — the product of media or press releases. Companies, individuals, and organizations create and/or report on news that is then disseminated to the various media deemed appropriate for the subject. This so-called "news" encompasses a wide spectrum of topics. Some of the more popular include:

- Legitimate news (new factory is built, new president elected, an industry first)
- Marketing news (sales records, market-share increases)
- Product news (new items or services offered to the market)
- Feature news (people stories, product-usage stories)

Information from situations described above is assembled in written format, supported with photos, videotape footage, or samples, and sent to the media in the form of a release. At this point, the solicitor loses control over the information (unlike paid commercials and advertising), and it is either accepted or rejected by the various media outlets solicited.

This kind of publicity will constitute the majority of your promotional campaign. Except for the time needed to create the information and any negligible costs involved — postage, phone, paper, envelopes, videotape — it's free.

Press Releases

The press release is the cheapest and easiest way to get your promotional campaign started. A one-page summary of the news you want to communicate, the press release has become the most overused, though still valuable, device in the industry. Distributed regularly to newspaper and magazine editors and broadcast news directors by the thousands every day, releases must be presented in the correct format to the correct person in order to work.

You want a release to be unique in content, not appearance. If yours does not fit traditional conventions, it usually hits the trash. Editors pouring through reams of information don't have time to decipher something non-standard. They want information in a familiar, convenient, and well-organized manner. Anything out of the ordinary gets ignored, or worse, posted in their office for co-workers' amusement.

Follow these press release guidelines to increase your chances for coverage in any solicited media:

Follow journalistic rules of form. Always double-space lines, leave a lot of white space, and double-check spelling and grammar. The same goes for any facts, dates, and names included in the release.

Don't write a book. Releases are one page — that's it. Keep rewriting it until you can capture all ideas in a succinct manner.

Give it a name. Use a short, descriptive, and attention-getting title on the top of the release. How else will an editor know what she is about to read? Try to make the headline offer or illustrate a benefit to the editor or reader. Make them care.

Don't forget the details. List company name, address, phone, e-mail, Web site, and a personal contact at the top of the page. Editors with questions need to know how to get ahold of you.

Let editors know when they can report the news. More often than not, a release will be immediately releasable, meaning whatever news is conveyed can immediately go to press. However, this isn't always the case. The inclusion of an "embargo date" is only used if you don't want the information published until a specific date. For example, you may have written a locally targeted release that reports on video stores in the area stocking your film or video. However, even though the sales have been made, it's possible there is some lag time (for duplication and packaging) before you actually get tapes delivered to the various locations. While it's valuable to have a newspaper story appear listing stores in the area that offer your movie for rent or sale, it can also be detrimental if the tapes aren't actually in the stores when the press hits the street. In this case, you might send the release with an embargo date for 10 days later, knowing you can make all deliveries within that time period.

Target your press releases. Of the hundreds of press releases received by media outlets daily, at least 10-25% have nothing to do with the publication to which they are sent. Read and become familiar with any magazine you are soliciting with releases. Do the same for any broadcast and Web-based media. Get to know what type of articles and press releases each specific media outlet uses and gear your information to the publication itself; don't expect that the editorial staff will take the time to critique, edit, and print your release.

Make sure your press release is news. This depends on the policies and practices of each specific media outlet. Your local weekly paper may be interested that you are staging a shoot involving 100 extras from the high school marching band, but the editor of *Independent Filmmaker* magazine doesn't care. He would be more intrigued by the fact that you successfully shot your film with a homemade Super 8mm camera, a fact unappealing to the local paper. Consider each specific media outlet and its audience when crafting a release. Remember, what is news to one is not news to all.

State the most important information first. Editors often only read the first line to sees if the release is important and appropriate. With the number of releases received on a daily basis, you're lucky to get this much attention. That's why your first lines must grab the editor's attention immediately. If that happens, chances are good that he will read it in its entirety. And this in turn makes it more likely for it to see print.

Answer the 5 Ws in the first sentence. Clearly point out the Who, What, When, Where, and Why of your release in the first sentence (or two).

These are the kind of answers an editor is trained to look for. Use the rest of the release to support this information.

Personally address the release to a specific person at the media outlet. This is where the research and surveying comes in handy. Hopefully you made some sort of personal connection when calling the various newspapers, magazines, and broadcasters applicable to your film or video news. Check your records to be sure you are sending news to the right person. Large newspapers and magazines have huge staffs, and an entertainment-themed release accidentally sent to the general news editor will not see print. Again, you have one shot to get things right. Also, don't address releases to a generic "editor." It only announces to that media outlet that you are too lazy to take the time to check details.

Call before sending. Make a very brief call (or e-mail) to any media contacts before sending a release. This alerts them to the information coming. After it's sent, do NOT call again to confirm receipt. You'll hear from them if they are interested.

Use quotes. Personalize and bring life to your release by using quotes whenever possible. For example, instead of writing "The video is a product of Jon Smith's more-than-fifteen years of Karate training," it's more lively to state, "I've studied the art of Karate for more than half of my life and put all my accumulated skills on display in this training tape." Quotes are more interesting than statements.

Don't get upset when you don't get print. More of your press releases will be ignored than used. That's not a reflection on your promotional skills or the merits of your film or video. It's a fact of physics. There's only so much space on a page or so many minutes in an hour that can be filled, and not every release that a media outlet receives can be used. NEVER call the media to ask why your release wasn't used.

Release Requisites
As pointed out above, a good press release answers any immediate questions a reporter may ask about a subject. Just that alone should get you some simple media mentions. But more detailed and thorough coverage is available, and can be had, if your release is cleverly crafted.

An example: You've shot a feature film in your hometown using local talent, crew, and locations. The movie's $25,000 budget was funded from so-called "creative financing" you secured after quitting your job six months before production began. When a release is distributed to local media relating these facts, it'll be tough for editors to ignore the "creative-financing" statement. This single phrase will elicit phone calls from the media who want to know what you did to raise that large amount of money in six months with no job.

This example raises another important point: Your job doesn't end when the release is sent. You better have a great story to back that "creative-financing" zinger. Some of the most publicized independent films received as much attention for their creative financing — credit cards funded *The Hollywood Shuffle*, Robert Rodriguez "rented" his body for pharmaceutical experiments to pay for *El Mariachi* — as they did for the movies themselves. In the example cited above, maybe your funds came from a Girl Scout-like cookie drive, by selling shares of movie "stock" to the locals, or simply going door-to-door asking for donations to make a film. Be sure that your backstory, supplemental information, background... whatever you call it, is the truth. It can be tempting to create a fabulous myth that would play well in the media and attract a lot of interest. But what will you do when someone exposes what really happened? Obviously, not all the promotional press that comes out of Hollywood is true. But those producers can afford to hire spin-doctors with the resources to handle damage control. As a self-distributor, you'll just lose credibility with the media, who will most likely never give you coverage again.

A successful release is one that causes greater interest on the part of the receiver — so much so, that he must contact you concerning the story. Since a press release is so easy to use, too many novice promoters mail them out to detail every insignificant aspect associated with a film or video project. I have a filmmaking friend who sent a release announcing the completion of his project's principal photography. This might be news if you're Francis Ford Coppola and you've been shooting *Apocalypse Now* for over two years. But when you've shot a low-budget movie on digital video in 23 days, it's nothing to brag about. As pointed out in the beginning of this section, press releases are the most overused promotional tool available. That's understandable since they're cheap and easy. Be sure when using a press release you have some news or interesting story to convey. Put yourself in the specific media outlet's position. Would you, as someone not personally connected to the film or video, find the information contained in the release interesting and suitable for your audience? If not, toss it. Don't abuse releases or use them as ego-bolsters. If you do, you'll find yourself unable to get any serious attention from reporters when you have real news.

Cut-Rate Communication
The best part of working with news releases is that you are able to reach a national audience for little more than the cost of postage, stationery, and copying. Depending on when you begin self-promoting your project, you'll want to continually develop and distribute press releases. Some ideas for legitimate release include:

- First day of location shooting (local market release)
- Call for extras
- Celebrity involvement
- Sales milestone
- Project available to various markets
- Use of new technology
- Honorable cause relation (a "Saving the Rainforest" documentary)
- Charitable benefit (portion of proceeds donated to charity)

Interviews
A very popular format of media coverage, the interview (witness the hundreds of television, radio, Web, and print versions of this tactic) presents you as an expert or interesting figure. For film and video-makers, the interview is a given. Rehashing a long night of shooting, telling a rags-to-riches story, or even spilling the beans on how you creatively funded a project are all stories that audiences (and the media) continually find interesting.

How do you get interviewed? Sometimes it is the result of a well-written press release. A reporter will call, wanting further information on whatever topic the release covered. Other times you'll "sell" yourself as an interview subject to the various media relevant to your film or video.

Mastering the art of the interview takes practice. A good starting point: local newspapers. Small-town media aren't usually looking for an angle of controversy and provide a good opportunity to "get your feet wet" with this public relations technique. Just remember that during an interview, you are the expert on the topic. It wouldn't be happening if you didn't have something educational, informative, interesting, or amusing to say.

Do your homework before any interview. Regardless if it's broadcast, print, or Web-based, find out the following:

- What is the background of the reporter? Is she familiar with film and videomaking, with the subject of your project, with the cast and crew, the locations, or with your personal life?
- Does the reporter ever look for "dirt"? Is she controversial?
- What is the editorial or "political" direction of the publication/station/Web site?
- What is the story angle? Is it to inform or entertain?
- What is the length of the interview? Will it last for 5 or 25 minutes?
- Where will the interview be conducted? At your home, shooting location, media outlet office, or studio?
- Is anyone else being interviewed (such as other independent film and videomakers) in conjunction with the story?
- When will the interview be aired or printed or made live on the Web?
- Who are the audiences? What do they want to hear?
- In what "environment" does the media present news? Is it humorous, newsy, instructional, or fluff?
- Will the interview be part of a series or a one-time feature?
- Can you get an advance copy, tape, or transcript? Usually this is unlikely, but you might as well ask.

In addition to learning all you can about the media outlet conducting the interview, be sure you really come off as an expert about the subject on which you'll be speaking. Know the answers to these questions:

- What was the news "hook" that generated interest in your story? Why was it chosen?

- What is your objective for the story? Do you want to entertain the audience, publicize your project, promote yourself, find money?
- What do you want the audience to know? What do you want to hide?
- What can you do to promote further interest in your project? What other ways can the audience get information?
- What are the answers to all the conceivable questions that could be asked?

You may want to ask for an advance list of the questions or topics to be covered during the interview, though not every reporter or editor will give this to you. Either way — be prepared. Outline on paper specific points or objectives you want to make concerning your film or video. Phrase these points as answers to potential questions and practice delivering the information verbally to a friend or into a tape recorder. Try to describe the production process with entertaining anecdotes. Practice talking clearly, directly, and concisely.

Many media interview situations exist; depending on the medium involved, each poses its own particular opportunities, benefits, and hazards. You'll want to take advantage of this publicity tool in as many occasions as possible, as it's probably the best way to get your face and project in front of an audience.

Radio and Television – Music and talk (paid and free) are the only things a radio station can rely on to fill airtime; the same can be said for the off-peak hours of many LPTV and local affiliate television stations. Network, syndicated, and paid programming fills only a certain amount of television airtime. The remainder is the stations' responsibility, and this is often scheduled with local/regional interview programming. (Haven't you ever watched TV on a Sunday morning?)

Radio presents one of the best interview outlets for the independent film and video promoter. Cable and low-power television networks are filled with interview opportunities as well — it's cheap, easy-to-produce programming. Thankfully, becoming a guest on one of these programs is fairly simple. Call the program director, or maybe even a disc jockey, and pitch yourself as a guest, your film or video as a topic, and tell them you'll be forwarding information. Send whoever is in charge of scheduling guests any materials you may have promoting your production. Also include a query letter that, in bullet-point fashion, outlines why the station's listeners would be interested in hearing what you have to say. Make sure your film or video sounds outrageous or appealing in some way. Just don't go overboard. Obviously, you don't want to fabricate a story. A little exaggeration is acceptable — and usually expected — to media representatives who've been working in the entertainment field. Lies, however, are not.

Chances are good, especially in the local radio market, the station programmer will contact you after receiving your information. If not, don't be shy about calling her back. Aggressiveness is not a bad thing when seeking publicity. Take this opportunity to demonstrate why you would prove to be a coherent, well-spoken, intelligent, and amusing guest. A DJ or program director will realize that how you present yourself on the phone is a pretty close to how you'll come off on the air — so be confident and direct with responses and speech.

Live media interviews are a two-sided coin. Sure the exposure is good, but unlike a print interview, where you have the chance to consider your answers before speaking and sometimes can retract what you said before the interview goes to press, a live situation doesn't allow for that luxury — not to mention the possibility of a raucous crowd of call-in listeners just waiting to heckle you.

To maintain consistency of responses, the person most familiar with all aspects of the film or video should try to handle all live interviews — unless, of course, the media specifically asks for someone else. You just want to avoid presenting multiple or conflicting stories about any aspect of the project. The best bet is to appoint (or hire) a **publicist**, someone who will be the first contact with any media request. That way, before any interview, you can meet and discuss possible questions, topics, and the appropriate responses.

Be sure to listen or watch any program before agreeing to be a guest. A personal example of why this should be done: My partner, our lead actor, and I appeared on a local cable movie program to promote *Pig*.

The movie is a serious drama about small-town police life. What we didn't know (since the cable network featuring the program was not available in our area) was that the movie program was done in a very tongue-in-cheek manner, and the host liked to play pranks on the guests and spoof their projects. Obviously, this was not a good forum for our film; the interview was very awkward since none of us wanted to play along. Had we watched the show beforehand, we probably would've declined the interview offer, but press is press (I guess).

Though radio interviews grossly outnumber television opportunities, getting your face on TV is still a possibility, especially since you are promoting a visual product. Make it clear you have footage available (in the station's needed format, too). I'm fairly certain most local and regional televised interview programs do not get the chance to showcase the work of independent film and videomakers as guests on a regular basis. Bring Hollywood to Heywood (or whatever city you might live in), selling the fact that you have a unique story to tell.

Whether it's radio or TV, remember the following when arranging and granting interviews:

- Interviews involve the human element — be prepared for anything.
- Mail a presentation package directly to the host of the program 2-3 days before the interview so he can be well acquainted with your project.
- Include a Q&A in the mailing to make things easy for the interviewer (and yourself). A list of potential interviewer questions, along with the appropriate answers, is a good way to start the ball rolling. Often, on-air personalities will read directly from this "script."
- Plan to arrive early. If the interview is live, and you're not there, that means dead air. Think you'll be invited back?
- Think before you say anything. Try to avoid too many "um's," "well's" and other non-communicative language.
- Microphones are very sensitive — try not to bump, tap, or cough directly into it.
- Hold a mock interview with a friend. Practice your answers and delivery to remedy any nervousness

- Never speak "off-the-record." Microphones are often "live" even when you think otherwise.
- Dress appropriately. You are a creative person. Wear what makes you feel comfortable; it'll make you more confident. Don't be a slob.
- If a question excites you, don't be afraid to show your enthusiasm. Don't feel obliged to be overly animated, but don't be boring, either.
- Maintain eye contact with the host or other guests when speaking to them.
- Be ready for the set-up of a radio or television studio. A DJ's booth can be small and cramped. Headphones are clumsy. Even small TV studios are filled with lights and equipment and people, making for a lot of distractions. Keep focused on the host and the questions.
- Take a promotional item along — a T-shirt, sticker, mug, poster, or promotional screener of your tape as a gift for the host.
- Ask the program's producer for an audio or video dub of the show. This makes an invaluable addition to your press kit.

As with all of your promotional efforts, follow-up is just as important as getting the press. Always send a note thanking those involved for the opportunity.

Magazines and Newspapers – Again, in the colossal market of magazines, weekly and daily newspapers, newsletters, and journals, print interviews are pretty similar to broadcast situations. The biggest difference: Magazine and newspaper editors usually contact you for an interview following a press release mailing. Reporting on a story, or a story in progress, usually involves getting quotes and comments. Writers will want to do more than simply copy the information you have sent them. This means a phone call and some questions, which is essentially an interview.

Whether you contact the publication, or they you, get familiar with every magazine, newspaper, and newsletter you solicit for publicity. This doesn't mean reading the last 10 issues cover-to-cover. What it does mean though, is knowing the difference between *Premiere's* editorial focus and that of *Filmmaker*. The former comprehensively covers the "Hollywoodization" of filmmaking — on set reports, interviews with big stars, upcoming blockbusters, and the like. The latter covers the independent market, with instructional and illuminating articles on how successful indies have "made it." To the uninitiated, these two magazines may appear, from their titles, as if they serve the same audience. But unless you acquaint yourself with a periodical's content, you'll never know — until you make the call and ask about interview possibilities. These kind of lazy mistakes are viewed as unprofessional and a waste of time. It's a quick route to finding the bad side of a potential PR contact, and could ruin your chances for coverage later, when your story may actually hold valid interest for the publication's readers.

Print interviews don't always require you to be "live." Though many take place over the phone, especially in the newspaper industry, a new trend is the e-mail interview, in which a writer will forward a list of questions, allowing you time to carefully consider each answer. The ability to think before speaking is the biggest benefit of print interviews. A live situation forces you to respond quickly to avoid "dead air" and looking dumb. Simple questions like, "What is the budget of the project?" can leave a filmmaker stumbling for words in a broadcast interview. Mentally fighting with issues like revealing the true budget versus a more inflated figure can get you tongue-tied. There's always one question you're never sure how to answer. With print, you can stammer, or be quiet for as long as you like, or even ask to come back to the question later in the interview. Nobody will "hear" you. It may not make for the happiest reporter, but at least you have the opportunity to answer questions as correctly as possible.

Just as the broadcast world creates "sound bites," sensationalistic short clips of words taken from a more complete and sensible answer, the print world produces short quotes from your longer, well-explained answers. Some refer to it as "misquoting," but the habit isn't always that severe. For example, you might describe your documentary film as "an artistic attempt to explore the beauty and terror involved in a homeless lifestyle." Change that to "the film explores the terror of homeless living," and the project sounds completely different from its true focus. Unless you are personal friends, never talk "off the record" or ask a reporter to "ignore" some aspect of your story. When doing an interview for *Pig*, a reporter asked us if the police departments helping with the production found the title offensive. I asked that we stop for a moment, and explained that when shooting with the various real-life police involved, we worked under the title *Small Town Cop*. To be honest, we weren't sure until the film was finished which title we'd use, but *Pig* won out due to its quick effect on people. The word instantly conjures up images and feelings, a great quality for a movie title. All of this information was told to the reporter under the assumption it was for his background knowledge only, not for publication. We also asked the reporter not to make any connections between our past films and this one. We wanted the movie to stand on its own, as it represented a big leap in scope, budget, and production value over past projects. Well, when the story ran, its headline read "Pig Out of Hiding," referencing and explaining in detail our naming dilemma. And the second paragraph described in great detail each of our past projects. So much for talking "off the record." Remember this example whenever you are tempted to say something you don't want in print.

Be prepared for your quotes to be shortened, your answers to be reordered, and the story to be presented in a manner that makes it most appealing to readers — which many times means making your words sound more sensational. It's just a part of getting media coverage, and luckily, the good examples outweigh the bad ones. Besides, what's the alternative? The old saying "bad press is better than no press" is very true to the independent film or videomaker looking to create publicity for self-distribution efforts.

Using Trailers

You'll be asked for or want to include a "product sample" in most media solicitation, interview, and event situations. In other words, the media wants to see a tape or DVD of your film or video. The problem is, supplying screeners to everyone who asks can get expensive. Dubbing, tape stock, packaging, and shipping costs add up fast. And there's no way of knowing who really "needs" a screener and who just "wants" one. Mailing a full-length copy of your flick for a new product listing in *Premiere* is worth any costs that may be incurred. Doing the same for a three-line mention in the local newspaper isn't. The solution? Send a trailer instead. Low-cost and quick to duplicate, trailers can, in most cases, accomplish the task of a full-length screener. Some quick trailer tips:

- KISS. Keep it short and sweet. It's a trailer, not a short story version of your project.
- Use a voice-over or music. Buy voice talent (local DJs work very cheaply outside of their on-air jobs) and hire a local musician to "score" your trailer. So-called "theme" songs and other audio cues really add a professional flair to an independent project.
- Make it "broadcastable." The visual qualities of your project are worthless if a broadcaster can't display them. Send trailers to TV and cable outlets in the same video formats they use for broadcast.
- Duplicate trailers on short-length tapes. These specialty tapes can cost as little as 50 cents if bought in quantities of 100 or more. It looks more professional to use the correct length tape for the project, instead of sending a two-hour VHS tape containing only three minutes of footage.

Video Press Kit (VPK)

This public relations tool combines interview, project, and supplemental footage to give the media a more complete picture of you and your film or video. A VPK should present legitimate content without trying to trick anyone about its basic promotional purpose. Essentially, as with a press release, you're trying to convey what sets your project apart from every other independent project out there.

Whether you use all original footage or incorporate clips from broadcast interviews and television/cable-show appearances, include as much as possible in the VPK that promotes you in a "rising-star" fashion. Just like a picture is worth a thousand words, a video is probably worth a couple million. It's easier and more fun to watch a 10-minute video than read through a 10-page press kit, and anytime you can make something more enjoyable, you're guaranteed a better response.

Like a full-length screener, VPKs are expensive and should only be used for prestigious print or broadcast media outlets. This kind of comprehensive visual presentation can help land you a bigger story and provides more material for broadcasters, so be sure to have the VPK available in the correct format for their use.

Running the gamut from half-hour programs to 5-minute promos, VPKs could include the following:

- Interviews with significant cast and crew
- "Behind-the-scenes" footage
- Talent screen tests
- Project trailer
- B-roll footage (sometimes known as "filler," this is a collection of generic shots of the production that can be used when a television-program host is talking)

Reviews

Probably the most recognizable form of "free" media available to the self-distributor is the review. Used by Hollywood studios and big-budget producers to both generate interest in a project and attest to its entertainment, instructional, or informational value, reviews can be used by independent film and videomakers in much the same way.

With the ever-increasing number of media outlets operating on the planet, the good news is that finding a source for reviews isn't too tough. A quick glance through any publication, or a flip through the channels, will uncover any number of review-oriented articles and programming. And, almost every single film and video advertisement — print or broadcast — features review quotes as part of its marketing message. Using reviews in this manner offers promoters a simple, cheap, and effective way to influence an audience. An objective, third-party endorsement of your project is always more convincing than saying the same things yourself — thus the popular habit of placing review quotes on film and video packaging.

While it would be great to have a quote from the *New York Times* or *Rolling Stone* heralding your project, that isn't too likely for those following the self-distribution route. Sure, there are exceptions — horror filmmaker Sam Raimi convinced Stephen King (not a recognized reviewer, but an obvious horror-literature icon) to publicly endorse *The Evil Dead*, and Roger Ebert has put his literary weight behind several extremely low-budget independent films. But those examples are few

---

Subject: "Vixen Highway": New Feature Tribute to Russ Meyer!
Date: Sun, 05 May 2002 22:02:03 -0500
From: John Adams Ervin <prolix@earthlink.net>
Organization: Berlin Productions
To: John Adams Ervin <prolix@earthlink.net>

Dear Editor:

I am an independent filmmaker who has just completed work on a digital
video feature that pays tribute to sexploitation maestro Russ Meyer.
Entitled "Vixen Highway", the story concerns three foxy female croupiers
who use
their sexy charms and physical strength to beat a string of lecherous
men in
driving an illegal donor liver to the house of an ailing rock star.

It would be an honor to have someone on your staff do a reveiw and/or
write-up on my feature and it's production, along with it's connection
to the wonderfully sleazy cinema of Russ Meyer. A VHS copy of the film
is available for viewing. You can also log onto the official web site
http://vixenhighway.com for production details, contact info and many,
many stills. I look forward to hearing from you and hope that "Vixen
Highway" will be featured someday in your publication!

John Ervin
Berlin Productions
Phone: (612) 362-5956

An example of a review solicitation letter sent to various media from the producers of Vixen Highway

and far between. The major entertainment and wide-audience general interest magazines and similar cable and broadcast shows have their hands full with Hollywood product. Stars sell magazines and get people to watch TV. The more people watch, the more these media charge for advertising. That's why the media covers "A" projects.

Studies have shown, however, that it's not the source of the review that is important, but rather what the review says. For example, the packaging for *Pig* features the quote, "Grabs hold of the viewer and never lets go." It can be hard to promote an independent drama like *Pig*, and that quote is a good, catchy line that really sells the feeling of the movie. Where did it come from? A small, syndicated radio review program on WHIZ, an NBC Radio affiliate in Muskingum, Ohio. Never heard of that source? Neither have most people. It still doesn't lessen the power of the words.

Not surprisingly, this technique of using quotes from lesser-known reviewers isn't solely practiced in the world of low-budget productions.

Often a Hollywood flick that went bad will feature a review from the *Ogden City Gazette*. The producers are well aware that submitting such a film to major print and broadcast outlets may only solicit negative reviews, so they work their way through smaller, regional media until they find one offering positive comments. Again, people are not interested in the source of the review — they just want to be told the film or video is good. As a side note, this technique has been especially beneficial to the once small-time reviewers who now have their names and comments splashed loudly on national print and television ad campaigns. Many of these once-unknown journalists have now made the jump to bigger and better gigs thanks in part to their positive reviews.

Another reason Hollywood turns to non-traditional, non-mainstream media for reviews is that film and video marketers have figured out the buying public now look to those same outlets as capable of providing trustworthy information and advice. As a potential moviegoer 15 years ago, you had *Siskel & Ebert*, *The New York Times*, *Newsweek*, and a couple of other so-called "respected" media offering reviews. Today, the Internet, specialty publishing, and digital cable have changed that set-up radically. *XtremeDays*, a teen-oriented movie featuring extreme-type sports, featured quotes on its marketing materials from *Thrasher* magazine. This publication is well read by members of the movie's audience profile and makes a great endorsement for the appeal of the project. If you want to reach your target viewers, you have to think like they do. Even if the *New York Times* championed *Xtreme Days*, it wouldn't have much marketing value to promoters trying to sell the movie.

Though often thought of as passé, reviews are still the most consistent form of marketing used to promote movies and videos, and their heavy use over time is responsible for two interesting anomalies associated with the technique. First, entertainment writers will often create positive reviews in hopes their words (and name) will get picked up for use in a project's marketing materials. Knowing a film's promotional campaign includes national television, magazines, and newspapers, it offers quite an incentive for a small-town hack to endorse a film that he may actually find horrible. Even well-known reviewers like Gene Shalit are guilty of creating very flowery prose that becomes irresistible to film promoters looking for the perfect quote. This kind of ego-press defeats the purpose and value of a review, but continues to happen on a daily basis. Pick up any paper and check out the movie ads for an example.

Another weird thing about reviews: Bad ones don't necessarily hurt you. The public, jaded to the review process (for the above reasons), realize that entertainment is subjective. Just because Joe Schmoe hates a movie doesn't mean all viewers will feel the same. Many movies open to mixed reviews where half of the national press love a flick and the rest think it should never be shown again. Who does the average viewer believe in that kind of a situation? Unless universal, a bad review is just more press, something every self-distributor wants. When beginning to sell *Killer Nerd*, we solicited a magazine called, appropriately enough, *Video Review*. The publication is no longer printed, but at the time was the most popular periodical available for anyone interested in the home video revolution. From equipment to movies, the magazine reviewed the merits of every product feeding the industry. Though not in the habit of reviewing independent films, the editors appreciated our initiative and thoroughness in providing materials, and let us know they would fairly judge the film in an upcoming issue. Needless to say, they hammered the movie, calling it "backyard trash." We were a little upset initially, but then found out the positive aspects of getting reviewed in a national magazine definitely outweighed the fact that the review was negative. Our little independent feature was reviewed on the same page as *The Terminator*, which obviously was a big seller and renter. Being featured side-by-side (including a photo of the video box art) to a multi-million-dollar Hollywood hit gave us a lot of legitimacy, making future sales and media coverage a lot more accessible. Plus, tens of thousands of people read that magazine, helping to increase general awareness of the project.

How do you get reviewed? The easiest way is to submit your project to one of the hundreds of print, broadcast, and Web-based media that exist solely for that purpose. It's best to check submission guidelines first, but most will want a full-length screener, any accompanying art, and other marketing materials and information. Be sure that you note where and through whom the film is available, as creating sales is one of the most important functions of a review. If soliciting multiple reviews simultaneously, be sure to keep track of release dates from all outlets that agree to cover your project. You don't want to miss clipping the review as an addition to your press kit. Most review-specific media are more than happy to take a look at your project and give it some press. It's what they live on.

General and special-interest magazines and other media that run review sections are hard to penetrate. With limited space devoted to the subject, they obviously are interested in high-profile projects familiar to their readers. This doesn't mean they don't cover an independent project now and then. Persistence, applicability, and professional presentation are the keys to this market. If a magazine like *Time* doesn't cover trashy horror films, don't waste your time and money sending them a screener for review. On the other hand, *Entertainment Weekly* has a multi-page section giving mini-reviews for new video releases, yet you rarely see a direct-to-video release featured there. Why? Probably because not many self-distributors have taken the time to solicit this publication for that very purpose. Always consider the audience for the project and what media they turn to for entertainment, education, and information. Those same media make good choices to solicit for reviews. Always call the editor or reviewer (if able) and be sure they'll accept your submission. Send materials with a short cover letter and indicate that you'll be calling back in a couple of weeks. Be sure to make the follow-up call. It's true that editors don't like being bugged, and I don't recommend follow-up calls with any other publicity mailing, but reviews are the exception. Quickly inquire if the film or video is scheduled for review, and if any additional information is needed. Plead your case if the contact is willing to listen. I've found reviewers to be much less time sensitive than other editors and reporters, as well as very film-friendly people. They're often up for a conversation, and this is the kind of personal connection that leads to coverage.

Solicit reviews at the final stages of your project, preferably prior to the creation of any final packaging. You'll want to use an especially good quote on packaging and marketing materials, just be sure to alert the media outlet it came from of your intentions. I've never come across a reviewer who wasn't agreeable to having his name and words on film and video packaging, posters, and advertising, but never assume anything.

Clip and copy all published reviews, print any Web-based reviews and — if possible — create an audio and video archive of televised and radio broadcast reviews. You will want copies of all media as it relates to your project for future use in your presentation packages.

USING PAID MEDIA

Unlike public relations, where your efforts produce uncontrollable coverage for basically no cost, paid media allows you to control exactly what you want to say — for a price. Widely known as advertising (though paid media does not always involve "ads"), buying coverage for your project has its advantages. There is no guessing as to what will be expressed in the coverage. You write and design the ad. There is also no guessing as to which media will cover your story. You buy the media of your choice. The big downside, of course, is cost. Just to put it in perspective, a full-page advertisement in *Premiere* magazine runs in excess of $50,000. That's for one ad, one time. That might exceed your entire production budget. At the other end of the spectrum, an ad on the *FilmThreat* Web site costs less than $200 — and you're reaching 90,000+ independent film fans. Which is the better buy? That depends on a variety of factors.

For paid commercials to be an effective means of persuasion, they must obviously be seen by those within your audience profile. Ads must also "position" your product in such a way that your target market is actually attracted to the film or video. It's not enough for potential audiences to simply "see" the advertisement, they need a reason to be drawn to it.

Most of this positioning work was done long ago when you developed a concept capsule and audience profile. Again, you will want to access that data to help with any creative and placement decisions you'll make in reference to paid advertising. Just as all of your promotional materials have contributed to the "image" of your product, so too will paid advertising. Actually, it is the most controllable way to position and create an image for your film or video.

The goal of effective media buying is to reach your defined target audience in the most efficient and economical way. To do this, you must consider a variety of factors:

Which Media Do You Use?

Large general-interest magazines and broadcasters are out of the question for self-distributors. You can't afford them so don't even dream about having your ad appear there. Fortunately, advertising is everywhere these days, so you still have a lot of options. Affordable paid media is available to the independent producer — anything from local newspapers and radio programming to regional cablecasters, from film and video industry trade magazines and specialty publications to flyers and placemats (yes, restaurant placemats). Consider your audience. What do they read, watch, listen to, or see on a daily basis? What else, besides the subject matter of your project, holds interest for them?

If your initial target audience is video retailers, you'll want to check out advertising in the video trade magazines. While not cheap, an ad in these publications can, for the proper projects, mean increased sales. Then again, you may want to reach directly to potential end-use buyers themselves. Maybe sponsoring a specialty cable television program makes more sense.

Materials for paid trade magazine advertising from the American Nightmare *promotional campaign*

To make educated decisions about which media to choose for a paid advertising campaign, you'll want to gather **media kits** from the various outlets available. These kits offer potential advertisers information on the media's demographics (their target audience and whom they reach), readership/viewership, size/duration of ads available, costs involved, and specials. Most media offer "agency discounts" of 15% or more to recognized ad agencies. What this means is that a $100 ad only costs $85 if an ad agency purchases it. The reason for this: A big part of an ad agency's business is called "media buying," where they create and execute a media plan. They obviously need to be paid for the time it takes to research and purchase advertising, and that 15% makes up that cost. Though you won't be working with an agency — they're too expensive — you can still get the discount. Tell the media rep with whom you are dealing that you are your own agency, making the buy for the film or video advertised. Some reps can get difficult at this point, but stand your ground. Make it known that is the only way you will buy the ad. Believe me, I have never encountered a salesman who would rather turn away advertising than bend the rules a little.

Another good item to study from media kits is an editorial calendar. This document lists the topics, features, stories, events, and news each specific media outlet will be covering at what point during the coming year. The editorial calendar allows you to advertise your product in an issue that provides the most supportive editorial environment. For example, let's say you have produced a video on motorcycle-restoration techniques. Reading the media kits from various motorcycle-enthusiast magazines, you find that in April, one of the magazines will be focusing on restoration, with articles and columns devoted to the various aspects of the craft. Thus, it would make the most sense to advertise your video in the April issue, since readers will be immersed in that topic. The same goes for feature films and the trade publications. Both *Video Business* and *Video Store* regularly devote the editorial in specific issues to Halloween-themed entertainment, exercise videos, independent and art-house films, and other genres of product. Buying an ad in one of these issues allows you to reach a more targeted audience within the targeted publication.

What Are Others' Experience with the Technique?

Talk to other independent producers and ask about their use of paid media. You can locate those practicing the technique by scouring the pages of magazines, Web sites, and other media that is appropriate to your audience.

Were films and videos sold with humor in the ads? Was it the classic "problem-solution" scenario (i.e., Overweight? Our video can help!)? Did the producers use a real-life testimonial from a satisfied client or a demonstration of the product? Or, was a plea made proclaiming the film or video's simple entertainment value? In addition to advertising technique, it's also interesting to note any promotions used in conjunction with the ad. These can take the form of a money-off coupon or rebate with purchase; a contest for cash, vacation, or other prizes; and the tried-and-true "buy one, get one free" so popular with advertising to the trade.

Possibly the most important consideration, depending on your film's genre and intended audience, is to question if anyone else is buying ads. If the answer is no, that might tell you something about the efficiency of the technique specific to your product. Or, it might illuminate the fact that everyone else is missing out on a good thing.

How Frequently and When Should You Advertise?

As strange as it sounds, you should advertise when business is heavy, not when you need sales. The theory is that more money is available during a good sales period, and you also have the ability to continue to influence present customers with positive-reinforcement advertising. If you wait until you need to sell some videos or DVDs, you are advertising out of desperation, hoping the technique will create business. It usually doesn't.

Obviously, any paid advertising should occur immediately prior to the release of your product to the market. That's when you want an audience to pay attention to what you have to offer — unless, of course, you have a product that is timeless in its appeal (an instructional tape, holiday, or seasonal-themed program). With these kinds of productions, a continually appearing ad may induce year-round sales. Any agency, publication, or person educated in advertising will tell you that a single ad is almost as useless as not advertising at all. The whole power of advertising is in repetitiveness. Or so they say. Sometimes a single ad placed immediately prior to the personal selling process works well in getting your project noticed. Sure, a two-month blitz of ads would do a better job, but you must balance that with...

What Can You Afford?

This is usually the first question that many self-distributors ask when considering an advertising program. More often than not, the paid-advertising decision hinges on budget, which may make the choice rather easy. Without a doubt, paid advertising lends legitimacy to a project and brings to it an immediate awareness level not easily duplicated through non-paid efforts. However, if funds are not available, it's an avenue you cannot explore at this time. Even if minimal funds are available for promotion, those monies might be better spent on direct, personal selling and public relations expenses. A comprehensive, yet low-cost PR campaign is certainly more beneficial to your film or video than a one-shot advertisement in a magazine. If a paid ad means sacrificing all of your other approaches to promotion and publicity — skip it.

Chapter 18 Summary Points

- The self-distributor must use as many available forms of media as possible for promoting a film or video project.

- Contact local press — newspapers and television, cable, and radio programs — for coverage throughout the production of a project.

- Trade publications — both entertainment- and subject-specific — present a valuable opportunity to self-distributors looking to score national media attention in a low-cost manner.

- Research the Internet, newsstands, libraries, yellow pages, and bookstores to find the proper media to solicit with news on self-distributed projects.

- To increase your probability of getting coverage, contact media outlets to check on submission policies before sending anything their way.

- A well-written and targeted press release is the easiest method of gaining "free" media.

- While interviews offer the film and videomaker excellent exposure, they can get tricky if you are unprepared.

- Never talk "off the record" to a reporter during an interview — there is no such thing.

- Enclose a broadcast-quality trailer or video press kit when soliciting television media.

- Self-distributors can use positive project reviews both to publicize their project and as part of any marketing materials.

- Surprisingly, bad reviews don't always hurt a project — getting the media coverage is the important part.

- When considering the use of paid media, pick those outlets that make the most sense for your project and your budget.

LOCATING, QUALIFYING, AND CONTACTING APPROPRIATE BUYERS

Self-distribution does not have to be a guerrilla-type affair. Sure, you'll be using "street-level" techniques to cut costs and find exposure, but you'll be doing these things in a very professional and business-like manner. While many independent filmmakers cling to their status as outsiders, this distinction is not so cool when it comes time to approaching potential customers. There may be some attraction on the part of a limited number of buyers to associate with a rebel "artiste," but most would rather conduct business as they always do. You'll want to mimic whatever ordering, shipping, and payment policies they have in place with other vendors. You want to make it easy for them to do business with you, because if you don't, they won't.

Regardless if you are working with rentailers, retailers, broadcast, cable, PPV, or institutional markets, the process of locating, identifying, and contacting potential buyers is virtually the same:

- Locate all possible buyers within the specific market category
- Contact these outlets to discover if they'll make for appropriate solicitations
- Create and distribute a mailing package
- Make call backs
- Negotiate sale

It may at first appear like a needlessly time-consuming process, but these five steps are essential for you to make a professional and worthwhile sales effort. You probably noticed that the actual "sale" is the last step. Why? Because you must be well-educated on every prospect you approach, ready for any possible query, and prepared with all required materials. As a self-distributor, you'll be battling against salesmen and agents whose whole careers have consisted of this practice. While you can't match their experience, you can level the playing field some when it comes to professional preparedness. That tactic may not get you a sale, but I'll guarantee it will garner you respect. And that is a very important quality to project when competing in this arena.

LOCATING BUYERS

Rentailers and Retailers – When selling to the retail and rental marketplaces without the services of a distributor, know on the front-end that you're in for some frustrating work. Not unlike your initial distributor hunt, the search for retail and rental buyers should utilize many of the same sources.

One of the first places you'll want to look: the local yellow pages — specifically, the sections titled "Video Tapes and Discs, Dealers" and "Video Tapes and Discs, Renting & Leasing." Depending on the size of your town, anywhere from two-to-200 or more shops will be listed here. Each offers great potential for buying your film or video. Why is this local market such a good starting point? As you hunt for buyers, you will — at the same time — be working the media. Those in your community, including any local retailers, will be aware of your project through stories in the newspaper and on local television. The press attention is nice, but the whole point is to use any awareness it generates to your advantage. That is why selling and promoting will be carried out concurrently. If you call a video store buyer offering your film the same day he reads about it in the newspaper, chances are very good you'll make a sale.

Selling to the local market also builds confidence. Response to your solicitation will definitely be more positive than when you begin cold-calling shops and chains across the country. Obviously, the project's hooks, packaging, quality, and pricing play a big hand in the equation. But local businesses like to support other local "businesses" like your own. You'll want to play up that fact.

Unfortunately, you can't just start calling the video stores and asking whoever answers the phone if they'd like to buy your movie or video. It doesn't work that way. At the least, you'll have a two-call process. First, call the store and ask for the name (and correct spelling) of the person responsible for buying product (usually called the video buyer, but small mom-and-pop shops may not be so formal). They'll ask why, so be honest. Tell them you are a local producer who has finished a project and you'd like to send them some sales information. I've had some bad experiences at this point, and I could never understand why. Those answering would tell me they couldn't give out the video buyer's name, or simply say they don't deal with local filmmakers and hang-up before I could utter another word. If you get a rude person on the phone, chalk it up to their loss. Be polite and thank all callers that allow you to. Start a database of all the store names, phone numbers, video buyers, and addresses you collect. Also make a note as to whether the store rents or sells tapes and DVDs or both. You will be using this information often during the callback process.

There are more than 15,000 video stores in the United States, so once you've covered the local scene, you have some work ahead of you. Finding independent video rental shops isn't the hard part — yellow page directories, online listings, mailing list companies, and industry trade organizations should produce a wealth of results. Keeping track of the data, especially throughout the survey period, may get tricky. The best advice I can offer: Take your time, be organized, take copious notes, and double-check every name and number before letting anyone off the line. As you look for buyers beyond the geographic boundaries of your community, attack one region at a time. You don't want to go hopping from state-to-state soliciting your film or video. It's just too confusing. Think of your strategy as war, with the entire United States in front of you as an area of conquest. You are one person; you do not have the resources to be everywhere at once.

Locate Regional Chains – The industry trade magazines are your best bet here. Both *Video Store* and *Video Business* publish issues that list the top retailers and rentailers (and various other buyers) each year. Again, start locally if possible. It really is just a matter of worming your way to the correct person in an organization. Always ask for the video buyer. All multi-store operations centralize their purchases through a single buyer who can track and respond to solicitations.

When looking for potential outlets, don't concentrate solely on video- or entertainment-product related businesses. You'll want to approach those stores first as they offer your best chance of sales, but don't fail to consider other retailers operating video/DVD rental and sales stations on their properties. Drug chains, groceries, gas stations, convenience stores, and other multi-shop retailers actively participate in video rental. Though many are supplied by wholesalers and rack jobbers, just as many operate their own video divisions. These retailers make great prospects for the independent, especially if you have a wide-interest project and low price.

Chains offer the self-distributor the quickest route to profit. Unlike working with a single store operation, which usually yields a unit or two sale at most, chains — for potentially the same amount of effort — can often mean orders for 100 or more tapes or DVDs at a time. Spend some time researching this market to be sure you've unearthed all potential customers. They're vital to your success.

Broadcast, Cable, and PPV Markets – For local television outlets, check your phone book. Whether a broadcast station, PBS provider, or cable network, they'll be listed under the headings "Television Stations and Broadcasting Companies" and "Cable Television Service." Nationally, you'll want to hit the Web. Several good sites offer "one-stop shopping" in terms of listing cable stations and networks, PPV providers, PBS information, and more. Specialty guides are also available exhaustively listing television outlets along with pertinent programming and contact information.

When contacting a potential broadcast-sales outlet, you'll want to speak with the programming director or acquisitions personnel if they have such a position. State who you are and immediately ask if they acquire independent product for broadcast. If they don't buy product in that manner, you want to know right away. For negative responses, try to discover where the network or station sources their programming. This lead could direct you to a sale. When reaching an interested party, you'll probably be asked to describe your project; if it fits their broadcasting guidelines (and is up to technical specs), they'll want a screener.

This is a wide-open and rapidly growing market. Cable is getting more and more audience-specific every day, so do your homework and hunt down every possible lead. During prime-time, PPV stations are heavy with Hollywood hits, but filled off-hours with "B" movies, providing a good market that pays well for the right projects. Leased access specializes in independently produced fare. And local PBS affiliates love to program "homegrown" productions. Whether you've made a feature-length film or a 27-minute instructional video, there are literally hundreds of sales possibilities with the televised market.

Institutional – There really is no "tried-and-true" method of efficiently reaching this market. Schools and universities, libraries, correctional institutions, daycare centers, churches, hospitals, and corporations are all very autonomous — meaning they do business the way they best see fit. Ordering and buying videos for members, clients, customers, inmates, and employees can be different with each group approached. This uncoordinated, non-regulated buying technique creates a lot of work for the self-distributor.

Your best bet with any institution is to call a main information line and ask for the person in charge of buying videos. This responsibility could fall to the president of the organization or a person specifically charged with the duty. Educational institutions and libraries often buy from "approved" vendors, which many times means a catalog service. If the buyer indicates such a practice, inquire as to the catalog company from which they purchase, and redirect you efforts in that direction.

Selling through catalogs is a whole market in itself, one that offers your product to a large number of potential buyers at almost no cost to you. Operating much like a distributor, the catalog advertises, sells, and processes orders for your film or video. You are then paid a percentage of each unit sold on a quarterly basis. The only downside to the arrangement is that you are responsible for creating a packaged product on the front-end.

As an example of the "anything-goes" operations inherent in institutional selling, many of those same outlets (like libraries) that only purchase from "approved vendors," will break the rules and buy direct from the producer. This practice is especially prevalent on a local level.

Contacting corporations and small businesses is a matter of product fit. Let's say you've produced a video concerning modern business-lunch etiquette. The tape, hosted by a well-traveled professor of business from your local college, explains acceptable present-day practices on what food and drink to order, how to eat, and how to handle payment and possible embarrassing restaurant situations. While you wouldn't consider the machine shop down the street a potential customer, you should call the 200-member law firm that just opened its doors downtown. Again, there are no real rules here on whom to call or how to get your foot in the door. If your project is something that would be of interest to employees, you might want to contact a corporation's human resource department. They deal with "benefits," and your film or video may be viewed in such a manner. Often, a call to the main switchboard followed by a short explanation of what you have and what you want to do is the best bet. It's not the quickest route, but eventually you'll find your way.

Finally, there are some independent distributors that specialize in selling product to the institutional marketplace. St. Louis, Missouri-based Swank Motion Pictures (*www.swank.com*), for example, lists colleges and universities, cruise lines, hospitals, motorcoaches, churches, and recreation departments as regular customers. Survey these kinds of companies as you would any other distributor, checking for needs, wants, audiences, and delivery requirements.

Foreign Markets – For the self-distributor, non-domestic markets are best approached through a foreign sales agent. You can contact and even broker with some foreign market outlets. Selling your project to British television is one such option. But the opportunities are limited, and without the experience and knowledge of how these deals (and the people making them) operate, the chances of "losing your shirt" are huge. Most foreign agents advertise and list their services in directories and online. Filmmaking and industry organizations (IFP, AIVF) are also a good source of information for contacts.

ACCUMULATE DATA, ACCUMULATE DATA, ACCUMULATE DATA

You want to gather as many leads as possible before you begin the solicitation process, because once that starts, you need to spend your energy on as limited number of tasks as possible. Let me explain.

Let's say you discover 10 local buyers that appear to be good prospects. If you start mailing packages to this group before you collect more names, then you'll have to continue surveying buyers, while at the same time calling on those already solicited — possibly resending materials to another group and closing sales with others. Don't get me wrong; there will never be a time when you'll be able to direct all your efforts to a single function. Self-distribution doesn't work that way. You will always be multi-tasking. Sales prospects will continually present themselves. Clients will need to be billed and re-billed. Tapes and DVDs will need to be replicated and packaged. There is a ton of work involved. But with some front-end preparation — such as collecting as many names as possible before you begin any other step — you can somewhat manage your task flow.

This situation does not mean you should gather data from *all* potential markets before selling. Rather, approach the universal market — one that includes every possible sales outlet for your project — in a submarket-by-submarket manner. For example, with *Pig*, we collected data on local single-store and chain video retailers and rentailers first. That group was our first target because we wanted, and needed to, raise some capital, and we knew local sales were a (mostly) sure bet. The process really depends on your needs and what you feel comfortable doing. Like any new venture, it's usually best to start small, moving slowly toward bigger conquests. Grouping geographically by business classification or territory exploited (domestic home video, foreign sales, DVD retailers) are options for dividing up the marketplace.

Chapter 19 Summary Points

- When looking for buyers, concentrate on one specific market category at a time to facilitate your efforts.

- Selling first to the local market allows novice self-distributors time to build confidence and experience.

- Chain-store solicitations by a self-distributor can mean sales of 100 or more tapes or DVDs at a time.

- Broadcast, cable, and pay-per-view present a number of lucrative, though limited and hard-to-access markets for the self-distributor.

- Though many sales opportunities exist with institutions, there are no standards for doing business with these entities, making it a time-intensive market to penetrate.

- Sales can only come through leads and information, so collect and accumulate as much as possible throughout the entire self-distribution process.

ADAPTING YOUR PRESENTATION PACKAGE

So, you've got this list of potential buyers. Now what? Well, it's time to revamp the presentation package used in your initial distributor solicitation, personalizing it for each market you approach. Your goal is to make your project as attractive as possible to each distinct prospect, using individualized hooks that appeal almost strictly to that buyer. Depending on what kind of outlet you are soliciting — independent distributor, broadcast/cable television, or video retail chain — the contents of each package will vary. Some will demand screeners, while others may simply need a sell sheet and contact number. Common sense combined with your market-survey information will determine who gets what. As you prepare the packages, try the following:

ADAPT THE SALES LETTER

In any local-market sales letters, note the number of people and businesses involved with the production. When we made *Pig*, more than 300 individuals and 20 local businesses were associated with the project in some way. We emphasized this "hometown" connection and the fact that we constantly communicated with our extended film "family," informing these folks where they could rent or purchase the film. In essence, we were telling video stores that the movie came with a built-in local audience.

Make a plea for support of local art. Libraries, gift shops, and other local, non-home entertainment retailers like to promote the works of local artists. Point out this fact in your sales letter. Our letter to regional library systems asked, "How many times do you have the opportunity to offer your customers a film made in the area by local artists?" A library is a community resource; promoting local talent is part of their mission. Make it hard for them to resist.

DEVELOP SPECIALIZED MARKETING MATERIALS

What good is it for a video store to buy a locally produced film if none of its customers know that the film was made in their backyards? Try simple, low-cost techniques to help store owners market your project. A round, neon-orange sticker was affixed to the front of every *Pig* package proclaiming the film was "made in Ohio." Four city names were listed around the perimeter of the sticker.

Luckily, the movie was shot in more than two-dozen locations around the state, so we could vary the cities featured on the sticker depending on the region we were soliciting. You want to make it as easy as possible for a potential buyer to say "yes"; any kind of marketing help you offer will get you closer to that outcome.

After realizing Borders Books and Music offered "local" book and music products, we created a counter display for the national media retailer.

Our goal was twofold: Have more of a presence than a tape on a shelf and prove that our film could be a top-seller even when merchandised beside Hollywood product in the megastore. The counter display was simply a blank cardboard VHS three-pack holder fitted with low-cost laser-printed graphics. Borders placed the display in their video department and quickly sold the initial three tapes as well as the nine restock units. After the success with the first location, *Pig* displays were sold to three additional area Borders and found the same success. Like the sticker, the display pointed to the film's local connections.

MAKE USE OF YOUR MEDIA

Remember all that media you've been generating? Now it's time to reap the rewards of your PR efforts. Whether it's a complimentary review, a feature on the project/people involved, or a "what's new" type mention in an industry publication, gather whatever media you feel is most persuasive and add to your presentation package. This compilation of clippings is assembled to exhibit the media's (and thus the general or concerned public's) awareness of your project. Many PR experts will tell you to include only stories from major media outlets when creating this part of the presentation package, as you don't want to "cheapen" your product by associating it with anything less than the biggest and best media. But again, the package should be solicitation-specific. A hometown newspaper article on your video probably won't mean much to a foreign sales agent. A *Playboy* review, however, will. For independent distributor solicitations, the more you can include, the more likely they'll "see" the potential for your project's wide-appeal success. It's a waste though to send a single-shop video rentailer more than one page of clips. They simply won't take the time to read or even look at an envelope brimming with paper. No real instructions other than "use your judgment" are needed here, just watch for the following:

- Make professional-looking photocopies of all media clippings. Use color if available.
- If possible, fill a page with a collection of good quotes or comments from a number of articles or reviews, along with their sources. People are more likely to read something if it's short and to-the-point.
- Put the big stories first and don't include too many. See above.
- Check every source's reprint rights policy. Most media will simply ask that you clearly identify the source of the material (the newspaper or magazine's name) on any copy.

PACKAGING

Self-distributors are responsible for product packaging — the colorful video sleeves and DVD cases, which, at times, can become the most important aspect of your presentation package (yes, sometimes even more than the actual film or video itself). Unlike acquired product, in which the distribution company handling the project will actually demand control over the creation of its packaging, the self-distributor must conceive and produce appropriate and attractive artwork that will be used to sell and display the movie or tape.

The Value of Art

Independent films, and many videos, live and die by their packaging. Major studio blockbusters benefit from the advertising during their theatrical runs — as well as the presence of stars, great locations, and amazing special effects — to draw in home entertainment renters and buyers. This is not the case with self-distributed product, which often has only one shot at catching a customer: its package. Even with a fair amount of media exposure, it's nearly impossible to compete with multi-million-dollar studio advertising campaigns, making the video/DVD package the first time a street-level customer becomes aware of the project. The packaging usually determines what step the consumer takes next. People are forced to make a decision from the information provided on the video sleeve or DVD case, leaving independent film and video selections the truest form of "judging a book by its cover." This fact is not lost on retailers and rentailers. Both are keenly aware of the importance of packaging and how its quality influences every purchase decision.

Independent distributors, video buyers, and anyone else who might buy your film or video to resell or rent it to the general public, want to see your packaging because it is part of the product. Good packaging is what moves your project through the maze-like distribution process. While first-level buyers (large independent distributors) are afforded the luxury of actually viewing your film or video before making a decision, this isn't the case with their customers. Subdistributors, retailers, and rental outlets must judge your project by the strength of the packaging alone.

Here's how it works: When an independent distributor agrees to handle your movie (meaning he'll sell it, but not take distribution "ownership" as in a traditional arrangement), or you begin the self-distribution process on your own, you or a distributor — depending on the contract — will create sell sheets. These sell sheets become the basic selling tool for the distribution effort. When the product is solicited to subdistributors and individual video retailers and rentailers, the sell sheet and any other promotional materials are used as a way to "advertise" the film or video to these audiences. Second- and third-level buyers must make a purchasing decision based on the information contained in the sell sheet, which is usually predominated by an image of the video packaging (see page 231).

Since they may be moving a large amount of product, all wholesalers, large retail and rental chains, and an occasional subdistributor will request (and should be provided) a screener. Retailers, however, are never given the opportunity to see a screener (at least in the self-distribution scenario) — it is just too expensive of a proposition. Could you imagine sending a screening copy of your project to everyone you solicit? You might as well just send them a clean dub and ask them to put it on their shelf. Regardless, small rental stores will still ask for screeners because they're used to getting them from the studios. Two reasons the big players participate in such a practice: 1) they can afford to, and 2) the studio is counting on the rental shop to order at least five copies (usually more) of the tape/DVD in question. As a self-distributor, you don't have that kind of clout, so tell a retailer the truth when he asks for a screener. It will kill some sales, but many video buyers — especially the independents — will understand.

Second- and third-level buyers are inundated with literally hundreds of sell sheets and promotional materials a month. What causes them to choose some films and videos over others, if the decision is based on artwork alone? That really depends on the buyer and the market he is trying to satisfy. As a general rule-of-thumb, big babes, big guns, and big names featured on packaging are elements most successful in drawing distributors and consumers alike. As trite as that sounds, it's the truth.

Video retailers, distributors, and other home entertainment buyers repeatedly report these elements — sex, violence, and star power — still work best in attracting people to product. Sure, a great story, quality information, polished production values, and superb acting help make a film or video worthwhile. But how do you translate these qualities visually on a video package or sell sheet? When video store customers decide on the attractiveness of an unknown film or video, the only information they are given to make this decision is found on the package they hold in their hands. With that in mind, it becomes easy to understand why the artwork representing your film or video is so important.

Packaging Pointers
On an independent, self-distributed project, it's going to be hard to get the "big names" as mentioned above. And while some films can feature

"big guns and babes" on their package, not every project lends itself to that kind of marketing. Like the product itself, the key to good packaging involves creativity. Consider your audience profile and concept capsule, as well as the following, when developing any artwork for your film or video:

Check out the competition. Go into a video store and really look around. What packaging draws your eye? What elements are found on these boxes that make them attractive? Take note of the different artistic and marketing approaches for the different genres. Use these same popular layouts and elements on your own package.

Use an easily identifiable image. For your key art (the image that represents your project), use an image that is immediately identifiable with your film. For example, with *Killer Nerd*, our key art consisted of a photo of the nerd (complete with leisure suit and bowtie), laughing maniacally in a blood-soaked room. Simple and to-the-point.

Use a solitary image. In a cohesive marketing campaign, your key art will be placed on everything that promotes your film. From the video box to posters to sell sheets, use of a single image makes it easier for people to remember the film. This is called **branding**. Like the golden arches of McDonalds or the apple on a Macintosh computer, branding literally does just that — brands your image on the consciousness of the public. It's no different for films. The egg used with the first *Alien*, the twisted airplane in *Airplane*, and the π symbol from *Pi* are all good examples of successful single-image branding on packaging.

Use a comedic or shocking photo. If possible, use shock value to attract attention. Sometimes the title alone can accomplish this (*I Spit on Your Grave* is not too subtle), but it works even better if backed up with a hard-to-ignore image. *Man Bites Dog* featured the main character aiming a gun directly at the viewer. This image has since been replicated countless number of times for other films. The *Dorf on Golf* series of videos always make use of actor Tim Conway in some goofy pose. The *Ernest* movies do the same with Jim Varney. It's hard to walk past this product on the shelf without taking a second look. Though regulations and laws exist prohibiting use of nudity and extreme gore on a publicly displayed package, look to the elements in your production that may lend themselves to this kind of exploitation.

Use bright colors. Understandably, black seems appropriate in some situations. But try to implement bright, neon colors and lettering in your designs. Try to imagine your box in a retail or rental atmosphere, on a shelf with 100 other tapes or DVDs. What will make it stand out?

Don't forget the spine. Sometimes your box will not be facing forward on the video store shelf. All that will be visible to the consumer is the side of the box, its **spine**. Be sure there's some graphic element and color on the spine — just in case that's what has to catch the buyer's eye.

No art is better than bad art. If you don't have the talent or resources available to create a professional looking package, don't do it. A bad first impression will make it hard for anyone to view your screener in a positive way, and you certainly don't want to feature an amateurish design on any promotional materials. Thanks to desktop publishing, quality graphic designers can be found everywhere. If you are really working on the cheap, try contacting area colleges and art schools advertising your design needs. These people are close to being professionals and will offer their services for next-to-nothing. Talk to professors or career-service staff about setting up an internship with one of their top students who would be responsible for a full-range of design projects. This kind of real-world experience (and wide range exposure) is very attractive to wannabe professional designers.

If discretionary funds are available for concept and creation of key art and packaging design, jump on the Internet and check out sites like *www.elance.com.* Here you can post the specifics of your project and get bids from designers around the world. All work, communication, and payments are done electronically, saving you time and money. Most of the artists working through such services are freelancing anyhow, so their fees are substantially less than what you would pay through a traditional design shop. Also check the local yellow pages for "Graphic Design." Many one-man operations will be listed there. Trumpet the fact that they'll be working on a high-profile film or video project that will get their design national exposure. Fame and notoriety motivates every kind of artist. Finally, if you must, cut a deal with a designer — exchanging his skills for some of the project's profits. Many other aspects of a film or video get made via the barter system, why not the packaging?

You almost always have to decide with what company you are going to print your video sleeves or DVD packaging before you begin the design process, because each firm has its own design specifics. If you physically pull a video box apart at its glue points, you can get a better idea of how the layout is constructed on a flat plane. Though not by much, this layout varies from printing company to printing company, and the difference of a tenth of an inch can make for a disastrous print job. Many printing firms offer downloadable design templates from their Web sites.

POINT-OF-PURCHASE MATERIALS

What are point-of-purchase (otherwise known as **P.O.P.**) materials? Anything used as an incentive to attract and convince any class of buyer to use your product at the "point-of-purchase." You see P.O.P. every time you enter a store. Be it posters and standees, interior signage and free give-aways, coupons and contests, P.O.P. is anything that can be created to persuade a customer to buy a specific product. The little coupon dispensers attached to a grocery store aisle offering discounts on the advertised product, the life-size Elvira beer display at Halloween, and that free toothbrush attached to the tube of toothpaste are all good examples of in-store P.O.P. But what if the customer's point-of-purchase is, like many of your potential buyers, wherever he opens his mail? Not a problem. P.O.P. can be created to meet the needs of any purchasing environment.

The whole point of P.O.P. is to give the buyer that extra push to choose your film or video over the competition. When you walk into a video store and spot an eye-catching poster advertising some video you are unfamiliar with, that's P.O.P. at work. There may be 10 new movies in the store that week, but you've already been impressed with the one.

The power of point-of-purchase materials is that they work through the distribution chain. If posters, counter cards, or other P.O.P. marketing materials convince consumers to buy or rent a tape, retailers want the P.O.P. to attract those customers — and distributors want (or create) the P.O.P. to attract retailers. In this way, P.O.P. satisfies everyone.

Trade vs. Consumer

Point-of-purchase materials can be categorized as one of two forms: trade-oriented — those materials beyond the screener used to attract distributors, video retailers, and other market buyers; and consumer-oriented — items that attract end-use customers at street-level.

Sell sheets, 8-1/2" x 11" full-color mini-posters featuring reproduction of the product's packaging, are used by distributors (including self-distributors) to convince subdistributors, retail chain, and individual store buyers to purchase a specific tape. Sell sheets also carry pricing information, pre-book and street dates. A pre-book date is the last day a tape can be ordered and the street date is the day the product is available or literally "hits the street." For example, a sell sheet may feature a June 5 pre-book date and a July 1 street date. The three-week difference allows the distributor time to create, package, and ship the number of units ordered by the June 5 date. Consumers are not usually exposed to trade-oriented point-of-purchase materials.

Sell sheet featuring the video-box art for American Nightmare. *Note pre-book and street dates in lower right.*

The Point of P.O.P.

Posters, bumper-stickers, give-away items, and other consumer-oriented P.O.P. are used to entice customers to rent or purchase a specific product. These items are usually supplied free-of-charge to retailers, or provided as an incentive to increase ordering levels. The counter displays for *Pig* were only supplied to those buyers purchasing six or more tapes at a time. We wanted to motivate local retailers to stock more than one or two copies of the film in their stores, and felt that an attractive (and free) counter display would draw customers to the product. Well-planned and professionally executed P.O.P. materials make it easier for the self-distributor to promote his product and find sales through the various retail and rental markets available. Conversely, ill-conceived and poorly produced marketing materials will turn off a potential customer. Following the same "no art is better than bad art" logic, be sure anything used to promote your project to the trade or end consumer doesn't come off looking like a 6th-grade craft project. Radical and edgy is one thing. Sloppy and cheap is something else.

In most situations, a project's distribution company will handle designing and creating any P.O.P. It knows the market, is familiar with buying patterns, and is tuned into what items might best be utilized with any given film or video genre. Additionally, as with packaging, a distributor will want to brand any product it represents as part of its "line," meaning point-of-purchase materials will feature recognizable characteristics of that company. It's no different for the self-distributor. In fact, you have more latitude with the type and variety of P.O.P. you can create because you are only concentrating on the promotion of one project. Large studios and independent distributors spend millions of dollars producing posters, **shelf-talkers** (those little cards that stick out toward the aisle from a shelf), and other point-of-purchase materials to advertise their latest releases. Though many retailers and rentailers appreciate the sales-support value of the materials, a number of this group also believe the promotions increase the cost of the product, and would rather buy a lower-priced video or DVD with no P.O.P. As a self-distributor, you can offer P.O.P. *and* low prices because of your limited overhead. And with a little-known, unadvertised independent film or video to promote, buyers may actually demand posters or some other point-of-purchase item just to familiarize their customers with the product.

Point-of-purchase ideas should be fun. When planning these kind of promotional items, let your imagination fly. Brainstorm! Write down everything you think of, regardless of how inappropriate or silly it may seem at first. Remember, you are marketing an independent film or video without the support of a large-scale advertising campaign. You need to attract notice, garner attention, and give buyers a reason to at least consider offering your film or video to their customers. P.O.P. is also valuable when generally promoting your project to the media or public in event situations.

In some instances, P.O.P. materials take on a life of their own and become another way to profit from a film or video. Familiarly known as **merchandising**, this technique allows for the simultaneous promotion and profit of a film or video from a product or service wholly separate from the film or video itself. Examples of this practice aren't hard to find. Children's movies lead the pack with merchandising deals. Toys, games, action figures, and clothing all provide promoters with opportunities to make money while publicizing a project.

Some Tricks of the Trade

Whether you are producing strict trade-related P.O.P. materials or a merchandisable product with sales potential of its own, the following hints will get you going:

Print posters and postcards. Probably the most important and popular form of P.O.P. is the movie poster. Whether it's billboard-sized or a mini version, the poster is most often a reproduction of the film or video's package art. Hundreds of firms design and print posters — it's just a matter of finding one that you can afford. Some quick tips: Print using four-color, glossy paper, unfolded or rolled, standard-size. You want the poster to be used, and if it doesn't look like all the others, it won't. Weird sizes that won't fit in display cases and cheaply reproduced posters will just get tossed in the trash.

Postcards are also handy. They offer the self-distributor a low-cost point-of-purchase item that can double as a quick communication device. Throw a handful in with each tape sold for retailers to use as customer premiums.

"Poignant ... a must see"
- The Dallas Morning News

"Lyrical and gentle ... a wonderful film"
- File Thirteen

"This is possibly the best teen misfit romance drama
I've seen since the mid-eighties."
- Ain't It Cool News

cicadas   mutiny productions presents

a film by kat candler

mutiny productions presents a kat candler film bryan chafin paul conrad
brandon howe don cass and introducing lindsay broockman director of photography jim eastburn
produced by scott bate kat candler and shawn higgins written directed and edited by kat candler
http://cicadas.home.texas.net

*Cicadas was initially
promoted to
distributors,
festivals, and other
potential buyers
through the use of
postcards.*

Make a soundtrack. Does your project feature original music or songs from unsigned bands? Create a mini-soundtrack (4-5 songs) on CD and either burn copies at home or through a low-cost replication service. Even small runs of 200 or less will cost only about $2.00 a piece, including packaging. A CD is a great addition to a presentation package, becomes a neat give-away for radio station interviews, and retailers will dig this kind of "collectible" gift they can offer customers. You may even want to sell any local media retailers on stocking the CD in addition to your video or DVD. Offering multiple products, or a product line, is always more attractive to potential retail sales outlets. Be sure the soundtrack packaging resembles the film/video packaging that has already been created.

The producers of The Dragon and the Hawk created a soundtrack, featuring both the film's score and "theme" song, as another way to promote the project to audiences and make it more attractive to potential distributors.

Create a printed guide. Are you marketing an instructional video? Is so, offer printed, step-by-step instructions that mimic the action on the tape. Package the booklet with the video and promote the product as a "complete instruction kit." Most people like the opportunity to "read" what they've seen onscreen, or refer to printed instructions, when learning a multi-step process such as cooking, deck-building, complicated dance routines, etc. Do-it-yourself home repair videos routinely come packaged with an ancillary set of printed instructions. Online, on-demand publishers offer very cheap prices for pamphlet creation and printing, allowing you to print only the number of copies you actually sell. Local print shops may be able to meet and even beat these mail-order prices, though you'll have to print a specific minimum quantity, which — even at low levels — may end up being a lot more than you need.

You can save money and avoid printing a guide altogether by using the Internet to "host" your instruction booklet. Much like the online tech support and manuals offered by software companies, an online guide can be made available to those viewers buying or renting your tape. A domain name and basic Web site can be purchased for less than $100 a year, allowing you to post multiple pages of instruction to correspond with the information presented visually. Users can access the site for more detailed instruction, and even print the pages if they want a hard copy. Feature the Web address prominently on product packaging and P.O.P. materials for maximum impact and use.

Books are not restricted to instructional videos. *The Art of The Matrix* was a highly successful text showcasing the artistic inspiration behind *The Matrix*, a mainstream feature film. *Slacker*, an independent and low-budget film from the early 1990s, was also supported with a book concerning its production.

Produce a "making-of" featurette. Was your project shot under extraneous conditions, at unique locales, or did it utilize eye-popping special effects? If so, you may want to create a making-of featurette to be packaged with the product. An example of this technique taken to the highest level is *The Matrix Revisited*, a two-hour plus documentary on the making of that blockbuster film. Probably at first conceived as a P.O.P. project to entice retailers and consumers, the project took on a life of its own and became a top-selling DVD both individually and as a twin-package with the original film.

Like a consumer-oriented electronic press kit, this supplementary video can include interviews, special effects secrets, and behind-the-scenes footage. It can be marketed to video retailers for in-store use (to play on monitors) or included as a supplement to the main product, allowing retailers and rentailers to offer it to customers. Due to a featurette's short running time, it can be duplicated cheaply and marketed with packaging similar to what is used for the main product. The P.O.P. product could also be tagged onto the end of a film. Though not a stand-alone piece, it still functions as an "extra" to draw the consumer to the film or video. DVD marketing has made an art form of this practice, including many hours of supplemental material on the discs. Promoters realized the format was very "collectible" and fed a hungry audience of viewers wanting more than simply a 90-minute movie. The habit is being copied in the video format, though not as successfully, because of technological space and random-access constraints.

Use photos as P.O.P. Select and duplicate several good production or "beauty" shots from your film or video for inclusion in sales packages. Photo houses will produce 8" x 10" color reproductions on cheap stock for as little as $1.00 a piece. On the dupes, have your star or "celebrity" write a personal note to each buyer you are soliciting, including an autograph, before sending them out. Many people collect movie stills and autographs, and who knows what the future holds for you and your cast?

Make stickers, pins, and other inexpensive novelty items. For less than $20, you can make 300 bumper stickers advertising your film. Divide an 8-1/2" x 11" sheet of paper into three equal parts. Print a message on each segment and hire a local printer to reproduce it on adhesive stock, cutting the sheets into three after printing. The same sheet could be divided into any number of configurations for different sized stickers. While you can't expect many people to actually place these on their vehicles, stickers are a cheap and easy way to dress up a large envelope, a presentation-package pocket folder, bulletin boards, clothing, and also make good "add-ons" for retailer sales. *Killer Nerd* stickers were included with every tape shipped, and many retailers placed them on their counters as a freebie for customers. Lapel pins serve the same function. Low cost and colorful (if well-conceived) pins can be quite popular, too.

Numerous catalogs and Web sites offering thousands of customized novelty items such as balloons, pens, notepads, mouse pads, watches, sunglasses, and more are available. Page through some, checking prices and product applicability to your film or video. Print your film or video logo on these low-cost items and offer to provide each retail buyer with a dozen or more as giveaways. Include in-store counter or shelf signs stating, "Free item with rental or purchase of this tape." While it might be a hassle for a big mass merchandiser, small video retailers love this kind of traffic generator.

Offer customized clothing items. T-shirts, hats, and bandanas, while not cheap, are wildly popular as giveaways. If you order in large quantities, t-shirts and basic baseball hats can be printed for less than $4.00 each. Include a t-shirt or ball cap free with specific quantity sales (i.e., one hat for every three DVDs ordered) in promotions to small, independent video retail shops owned by folks who dig these kinds of things.

Market the main character. This isn't limited to mega-million-dollar Hollywood films. *The Toxic Avenger* and *Attack of the Killer Tomatoes* both started as extremely low-budget movies that have now spun off into toys, comic books, and cartoons. Is your main character a marketable entity? Create a merchandising or P.O.P. idea around that idea. Even slasher films are marketable in this way: Witness Freddy Kruger and Jason toys and merchandise. This idea is not reserved solely for feature films. Many children's and educational videos feature a main character that has become a merchandising phenomenon (i.e., *Bob the Builder*).

Use a "piece" of the project as P.O.P. This technique depends on the subject matter of your film or video. Horror film producers might want to package a signed, certified, and sealed card containing a drop of the star's blood along with any tapes sold. A nature documentary might be well promoted by the inclusion of tree and plant seeds.

Encourage viewer participation. Create a fan club for viewers to maintain project awareness. Offer "free" membership if the viewer watches the tape and responds to a phone number, mailing address, or Web site posted at the end of the program. Send respondents regular newsletters and offers to buy any merchandise you may have created to promote the film. Use a contest to drive rentals and sales. Offer an appropriate prize (free one-month gym membership for exercise videos, tools for home repair instructional tapes) to contestants meeting some sort of purchase criteria.

P.O.P. and Sponsorships

Self-distributors should also investigate any co-op or sponsorship opportunities that may exist with local, regional, and national corporations that could be aligned with a film or video. For example, let's say you produced a deck-building instructional tape. A national tool manufacturer could get involved with this kind of project in several ways. First, the firm might provide 200 specialty deck-building hammers in exchange for its name being featured on the video package. As you solicit video buyers, you can offer the hammers to this group for specific quantity purchases or create a contest for the final consumer. The tool manufacturer could also **sponsor** the video. In this arrangement, the tool company buys the video from you as a give-away to their customers buying the hammers. Since the hammer is specifically designed for deck building, packaging the product with an instructional tape on the subject is a natural. Of course, a sponsorship deal works best if you arrange it prior to production (so the same hammer could actually be used by the instructors in the video), but it's not necessary. Be creative when considering sponsorship opportunities. Exploit what you have. Look at your film or video, its hooks, stars, locations, story, effects — anything associated with the production that will allow you to promote it in a unique way. As a self-promoter you need to scream for attention by outsmarting the competition. Anybody can spend a lot of money promoting a project, but it's no guarantee for success. A good idea will always get more notice.

Chapter 20 Summary Points

■ Try to make your presentation package as personalized as possible for each potential buyer.

■ Packaging is often a self-distributor's only chance to attract the attention of a customer.

■ While creativity is always important in the design of packaging, check out the competition and work with a professional artist to ensure the best results.

■ Sell sheets are the most powerful, useful, and affordable point-of-purchase tool for the self-distributor.

■ The greatest value of point-of-purchase materials is their ability to attract attention to self-distributed films and videos that, unlike their studio-released competition, don't enjoy the benefit of a multi-million-dollar ad campaign.

■ A singular, identifiable image — one closely associated with a film or video — must be the focus of every point-of-purchase item.

EXPLOITING OTHER PROMOTIONAL OPTIONS

Immediately before and during the sales process, you'll want to examine the possibility of using a variety of other promotional techniques at your disposal. These tactics will help you raise awareness for your project at a very crucial point of the self-distribution process: when you are soliciting potential buyers. The more attention you can gain when making sales, the more likely you'll achieve positive responses. At first, some of these techniques may seem inappropriate for your project — a premiere for an instructional deck-building tape? But the more you think about your film or video in a promotional context, the more suitable each of the following tactics will become — premiering that deck-building video to a group of volunteers about to construct a Habitat for Humanity project, sparking other ideas of how to promote your film or video in a non-traditional manner.

PUBLICITY STUNTS

The purpose of a publicity stunt, which can be presented in a serious, amusing, or outrageous manner, is to provoke interest in both the buying public and the media — in a unique way. It's usually not so much something you want to invite these groups to witness (though that's not a bad idea), but rather something you do that will create attention on its own.

Take a cue from Hollywood and get the media together for a set visit. Though in strictest terms, this is not a publicity stunt, these "behind-the-scenes" media **junkets** (a fancy word for working vacations) are a very expressive way to communicate your story to the press. The "set" can be anywhere the actual production is shooting. Obviously, the junket has to take place during the production phase of your project, but even if your film or video is "in the can," trot out the gear for a "reshoot" — the attending publicity is worth the trouble.

Though old hat to most national entertainment reporters, a set visit is a pretty exciting proposal to any local/regional broadcasters and print journalists. Pump up the fact that it is a one-of-a-kind experience, offering unique photo, video, and story possibilities. Make this press day a true media event by:

- **Pampering the invitees.** Have snacks, meals, and plenty of drinks on hand; provide transportation, cover from the elements, and a small promotional item like a hat, shirt, or key ring (just don't call it a gift, because the media can't accept those).
- **Making it worthwhile — for you and them.** Bringing out the media requires time and effort and expense. Shoot something exciting to thrill those attending, then don't be embarrassed to ask any reporters and photographers when you can expect to see the story in their respective media outlets.
- **Providing exhaustive information.** Make sure everyone is given a synopsis of the project; fact sheets on cast, crew, and locations; and a schedule of events for the day.
- **Babysitting the media.** Get someone you trust to personally "handhold" the group of reporters, basically granting their every wish for access, information, food, and bathroom breaks.
- **Following up.** Send a "wrap-up" press release detailing the event as well as a thank-you note to everyone that participated.

Drawing a Crowd

A more attention-getting publicity stunt is one involving the public — the best scenario being unsuspecting bystanders in a public area. Referring to the fore-mentioned documentary on lake pollution, on the day of your first public screening, hand out water bottles filled with "polluted" water (actually clean water, colored brown) with labels stating "water courtesy of your friendly neighborhood water polluter." While you may feel the ire of the actual companies accused of contaminating the community's water supplies, everyone handed a bottle will indeed take notice of the display.

If your horror film features special gore effects make-up or alien costumes, put your actors "in character" and parade them throughout the city, distributing flyers announcing the film. I'm pretty sure a man strolling down the street with his arm hacked off will grab the attention of everyone on the block.

Creating Controversy

Confrontational settings also make for successful publicity stunts; if your film or video features a controversial topic, exploit it for all it's worth. Perhaps you've shot the independent version of *Charlie's Angels*, twisting the storyline so the girls are a little less resourceful and a little more risqué. Well, chances are the local chapter of the National Organization of Women (NOW) won't be all that happy with your depiction of the fairer gender. Yet this group might not even know about your film. Send them a screener and invite them to voice their views. Or, better yet, act as an anonymous informant filling them in on the content of the film, as well as where and when would be a good place for a protest (the premiere). Pickets, strikes, and boycotts are great visual publicity stunts that create attention and get the media running.

Stunt Stumping

But what if your project seems to contain no controversial topics? All films and videos do — you just have to find them. While *Killer Nerd* was basically a horror comedy with no blatant controversial elements, we were able to exploit one "angle" to create publicity during the premiere. This angle came from a story in the newspaper about Harvard college students starting a "Geek Society" to lobby for longer quiet hours for studying in the dorms. From that idea came the "American Geek and Nerd Society," an invented group created for the sole purpose of protesting the manner in which a nerd was portrayed in the film. Six hired actors were dressed as nerds (short sleeve dress shirts, polyester pants, glasses) and supplied with picket signs and flyers. They marched in front of the premiere hall, hassling those waiting in line. We even went so far as to rent a post-office box under the Society's fictitious name and have the address printed on flyers in case anyone in the media wanted to check the group's authenticity. All the actors involved were coached with fake names and answers for possible questions posed by bystanders and reporters. This stunt was highly successful. It landed us on MTV, CNN, and dozens of affiliate and local news programs and print media.

Examine your film or video for elements that you can exploit through a publicity stunt. This tactic is necessary for self-distributors because you do not have the budget available to advertise your project through common means. Exploitation of a project does not always have to carry bad connotations. For that documentary on polluted water, stage a premiere at the lake's public access area. Use a DC-powered TV/VCR and screen your project in the parking lot (with permission of the park authority, of course). Bring blankets and folding chairs for people to sit on and invite an ecology expert as guest speaker to talk after the screening. Not only will you receive tons of free press from the event, you will also draw attention to the potentially harmful effects of pollution on the environment and to citizens using the lake.

As the preceding example demonstrates, publicity stunts are limited only by your imagination. Some ideas and cautions to keep in mind when utilizing this promotional technique:

- Make a float for local parades. It doesn't have to be fancy. If you've produced an instructional dance video, have your hoofers dance their way through the parade route, distributing flyers and handouts promoting the tape and its sales locations. *Killer Nerd* was publicized via a local parade. The star of the film rode in the bed of a pick-up truck, throwing packets of Nerds® candy to onlookers and passing out flyers that highlighted stores in the area renting and selling the movie. At the end of the ride, he signed autographs for fans and mingled with the crowd. The stunt was unique, drew a lot of attention, won accolades from store owners who purchased tapes from us, and only cost about $30 for the candy and flyers.
- Create a human billboard for your film or video. Place actors in costume, displaying a sign advertising the project, near a busy pedestrian or traffic intersection.
- Stage appearances at local video shops. The MTV-VJ star of *Killer Nerd* did guest appearances at area video stores that purchased multiple copies of the tape. He signed autographs and posed for pictures on several Friday evenings — a busy time for video rental stores. The stunt drew a lot of customers and pleased store owners with high tape-rental activity.
- Sponsor some kind of contest that creates both consumer interaction and visual interest. Look for "participatory hooks" in your project — those things that allow for a natural association with a contest. Writing, songs, costumes, physical endurance,

memory, strength, dancing, and athletic ability are just some of the many topics on which a contest can be based, and each allows for a very public display.

■ Avoid embarrassment. Three reasons why a publicity stunt can go sour: arrests, lawsuits, and no-shows. Skirt these situations by operating within the limits of any laws that govern such events, making all participants sign releases covering any personal property or other damages, and tipping off the media to your plans if necessary.

Possibly the biggest "stunt" you can pull is a premiere — the first-ever public showing of your film or video. Since this event usually guarantees extensive press coverage (you actually invite reporters to attend, as well as a hundred or so other folks), you want to put on the best show possible.

PREMIERES AND SPONSORS

Imagine the following scenario. You've worked for months, maybe longer, promoting and selling your project. Hype surrounding your film or video is at a fever pitch. Your project has been the focus of numerous newspaper and magazine articles, and in the past week you personally have appeared on radio and television to promote your creation. There is a publicity stunt taking place outside of the premiere hall, posters advertising the production are on every corner, and an enthusiastic crowd of people are lined up around the block waiting to get in.

This is a best-case scenario for a premiere, which, compared to the other promotional techniques in this text, is going to be expensive. Obviously, if you had the forethought to budget for marketing and promotion, you would have funds allotted for such an event. But if your project is finished and you are now trying to attract media and distribution offers through this technique, chances are good that all you can afford is rolling out the TV and VCR — and possibly a bag or two of Orville Redenbacher's popcorn — for a "showing" in your family room. Not too impressive. Luckily, even without a bagful of cash, an impressive premiere is possible. You just have to work with other people's money.

To illustrate the point, let's think big. The plan is to rent an actual movie theatre and mimic the latest Hollywood opening — you know, red carpets, bright lights, and limos. The first question is often "Where do I get all of these items, especially in Podunkville?" Contact a wedding planner to start. If she can't find it, she'll know where to send you. Question number two then becomes "How much is all of this going to cost?" Realistically, like a wedding, you could be looking at upwards of $5,000 or more. A line-by-line budget can be created after consulting with the various planning and service providers.

While it's good to have a premiere budget target in mind, you'll want to have some back-up options ready, as you may find yourself working backwards. You might have to plan your premiere around how much money you "find," and go forward from there. Realize on the front end of this process that nothing is set in stone, so compromises to the original plan won't be disappointing.

Other People's Money

Working with other people's money quite simply means you want to find sponsors to cover the costs of the show. If you were helming the release of a Hollywood production, it would be no problem to call your friendly Pepsi, Blockbuster, Gap, or any other major consumer-product advertising rep and inform them of available sponsorship opportunities. These big companies would swoop in, erect large signs, provide refreshments and goody bags, and foot the bill for the whole night in exchange for the ancillary publicity they would receive from their sponsorship. Obviously people attend premieres to see a movie, but there's a lot of downtime associated with such an event, and advertisers capitalize on that downtime by putting their names, products, and messages in the faces and hands of everyone attending. If sponsors can get patrons to take something home sporting their brand name, all the better. And this obvious "logo-izing" of the event means those same names will appear in the backgrounds of any media photographs or videotape.

However, you're operating in the world of independent, low-budget films and videos, and the cold truth is there will be no major corporations lining up at your door to throw in some bucks for your production's public release party. Can you blame them? Why would an internationally known company associate itself with a project that probably cost less than the catering for their last commercial shoot? To be fair, these firms do get involved with high-profile film festivals and some independent projects that hold national appeal. Once in a while a large corporation may back an environmental piece or a thoughtful documentary on some aspect of our culture (Native American life, changing family values, advances for the handicapped, etc.). There are also some limited opportunities for very commercially appealing projects that target the same audiences as a potential sponsor (clothing manufacturers for teen-oriented flicks is one example), but only those produced at a very high level of sophistication and distributed on a national basis. Usually with independent fictional features and videos, you'll find closed doors to your premiere proposals. If you do find yourself soliciting promotional support from a nationally recognized company, get ready to journey a very long legal and bureaucratic path.

Local businesses within your area present better potential as premiere sponsors. Owners and employees may already be familiar with you and your project, making it easier for your solicitation efforts. Plus, hometown shops have the most to gain from any secondary publicity surrounding the event, as most premiere attendees likely live, shop, and do business in that area.

Back to that $5000 price tag. Working with sponsors means you are going to have to chip away at that number until there's nothing left — or even better, there's a surplus. Luckily for us creative types, this isn't rocket science. What you need to do is offer a potential sponsor something in return for his monetary donation. The more money someone gives you, the more they get. And vice versa. You want to create a "menu" of sorts that outlines all the sponsorship possibilities attached to the premiere. These can include:

- Business name on tickets
- Banners bearing business name hung in theatre area
- Business name added to the film before credits (i.e., Marvin's Motorcars presents...)
- Business commercial played before movie (this started with *Top Gun* and Pepsi)
- Business name on posters and any advertising
- Business name on programs, t-shirts, and any other hand-outs

- Business name and link on film/premiere Web site
- Business name on video/DVD packaging
- Business product/location featured in production (i.e., shoot a scene at a sponsor's bar — if you planned early enough)

You get the idea. Basically, you want to sell any "space" you have available. To do so, approach sponsors in a professional manner. The process is very similar to hunting for distributors. You need to research the market to determine which companies "fit" the occasion, identify the decision-maker, place an introductory call, send a presentation package (sales letter, sponsorship plan, press clippings, trailer, and any other marketing materials you've developed to promote your film or video), and follow-up after a couple of weeks. Make it hard for people to refuse by offering a wide range of involvement levels, so even the poorest can get in on the action.

There's Always Plan B

Remember that best-case scenario outlined earlier? You know, the one where your premiere rivaled that of the latest Vin Diesel flick? Well, let's say after several rigorous and what seemed like "shoe-in" solicitations, you still haven't struck gold. Now what? Well, you're gonna have to dig into your bag of promotional tricks (you do have one, right?) and come up with one of two things:

- A better way to ask for money
- An alternative plan for the big night

For those in the "won't-take-defeat-quietly" crowd, you might want to re-examine your sales skills. How are you approaching potential sponsors? Are you making them feel like they are getting in on an exclusive event? Do they think there is limited space to participate? Are they aware that other similar, competing businesses are already part of the event? Are you really offering them something of value? Is there a true connection between those you are soliciting and the project? If you are doing none or few of the above, you may want to try changing your "plea" to reflect some of those ideas.

However, sometimes a tepid response to your premiere is not an indication of your abilities as a salesperson, but just a reflection of interest in such an event. There is a good chance that because you are probably an unknown entity (meaning, you can't compete with well-known and trusted civic organizations that are often searching for event sponsors, too), nobody wants to bite. If this is the case, it's time to consider another way to publicly debut your film or video.

And Even Plans C,D,E...

Like the saying goes, you can't drink champagne on a beer budget, so it is time to downsize. Maybe, in lieu of the movie theatre, you can hold the screening in a less-expensive university or high school auditorium. Go for video projection instead of making a film print. Roll up those red carpets, save electricity by turning out the searchlights, and take the bus instead of the stretch towncar. Obviously, you don't have to go to these extremes, but you need to consider all aspects of the original plan, looking for areas to cut costs. One caution: Don't "cheapen" the night too much. Sometimes it's better to cancel than to try to pull-off something you cannot afford to do right. We've all been to cheesy weddings — it's the same idea here. Compromising your original plans can also take all the fun out of the premiere, so keep that in mind if you are faced with this dilemma.

A second way out of the cash-crunch situation is to collect "in kind" sponsorships. What this means is that instead of a cash donation, a company or individual donates time, facilities, or services to your premiere. So, instead of needing a couple hundred bucks to print tickets and programs, the printer donates his services (and paper) to the premiere, in exchange for placing his business name on every printed item. This technique works just like cash, except that every "sponsor" you work with has to offer something appropriate to your big night. Some ideas: limousine services, caterers, bars, theatres, party halls, tuxedo companies, party suppliers, and even security firms.

That's a great way to get the gig going, but what about promotion of the event through the media? How can you afford a newspaper or cable television ad without any hard cash on hand? Well, most major regional media will not provide services (their "space") in kind, unless it's for a benefit of greater import than making you a star. In other words, you are going to have to find a worthy (and appropriate) cause that your premiere can benefit. This means charging a modest fee for admission, so you have something to donate to the charity, organization, or individual bene-fiting from the event. Try to find a natural link between your film or video project and the many charities in business. Initial plans for a *Pig* premiere found us allying with the local Fraternal Order of Police (F.O.P.). Fortunately, local police departments understood the title of the film was not meant in a defamatory manner, but rather used to hook an audience, and were very interested in being part of the premiere. Ticket sales would benefit the groups' annual "Shop with a Cop" Christmas events. The union of the film and the police group not only made sense, but also helped draw a lot of media attention to the event.

Some other ideas to make your premiere successful:

- Solicit smaller local broadcasters and print media to become sponsors. Large media will want you to pay them for this kind of an arrangement, but cablecasters, LPTV, and low power radio stations rarely get this kind of an opportunity. They can pass out branded goodies to the crowd, bring in their equipment for a live remote broadcast from the premiere, and talk up the event and give away tickets for several days preceding the event.
- Stage a publicity event on the night of the premiere (very good idea to maximize publicity), and be sure the press is there to cover it. Encourage friends and possibly the manager/owner of the premiere facility to place similar calls, but don't overdo it.
- Invite all area video retailers and rentailers by contacting your nearest VSDA chapter. Also contact any regional sales reps from the large national distribution companies (Ingram, Baker & Taylor, etc.). It's good to meet and network with these peo-ple as they can give you valuable feedback — you might even get offered a distribution deal!
- Hold the premiere in a place that is fitting for the project. A bar is a good place for a "party" film, a college auditorium for an edgy independent movie, a local library screening room for an educa-tional or children's video, a health club for an exercise instruc-tional tape… the choices for a venue are virtually limitless.
- For "issue-oriented" projects (documentaries are a natural) don't be afraid to invite those with opposing views on the sub-ject. Their attendance can make for an interesting, lively, and media-attractive give-and-take.

- Don't forget the basics: Personally introduce the film before it screens, provide food and drink (possibly specific to the project theme, i.e., sports drinks and protein bars for an exercise video); rope off a private media seating area; secure the services of a "celebrity" host; consider security, parking, and bathroom needs; and don't forget to set out a table to sell copies of your project. At first, the latter may seem tacky, but it is the absolute best time to sell your product. Those attending the premiere will be caught up in the enthusiasm of the film or video, and be very willing to show their support. Think of the last time you attended a concert and walked past all those guys selling t-shirts and programs. Same idea.

FILM FESTIVALS

Deserving of and extensively covered by a slew of books and articles specifically dealing with the subject, film festivals offer the self-distributor another promotional option — beyond traditional advertising, media, and personal selling — to publicize her project to potential buyers.

Originally designed as a forum to present new and original independent films to a more "art-minded" audience, film festivals have now become sales markets and publicity stops for "almost mainstream" movies aimed at the general public. While the big festivals like Sundance, Slamdance, Toronto, and Telluride can really launch an independent film, the politics and networking involved behind the scenes is really out of reach to most independent self-distributors. So what's left? A lot.

The Wind was extensively promoted via the film-festival circuit, where it won several awards. This festival recognition was later used in the film's promotional materials.

Literally hundreds of festivals are occurring around the globe throughout the year. Not a month (or even a week) goes by without some volunteer projectionist unspooling the latest independent flick in front of an enthusiastic group of film-savvy viewers. Showcasing shorts, features, experimental, 8mm, digital video, animation, adult, documentary, music, and even pixelvision movies, there really is a festival for everyone. But does that mean you should enter? That's a tough question — the answer depends on each self-distributor's specific goals, needs, and situation. Before you start filling out entry forms and writing checks, consider the following to decide whether film festivals will benefit or detract from your promotional efforts:

How Much Will the Film Festival Cost You?
It's not cheap entering a film festival. Sure, many have "no entrance fee" regulations, but those offering the most potential visibility are never free. Usually lightening your wallet anywhere from $50 to $75 or more, these festivals involve other costs beyond the official fee. You'll need a good quality screener, media kit, postage, and packaging materials. Enter 12 festivals at $75 a pop, plus postage and materials costs, and you've spent nearly a grand — with really nothing to show for it. If your project is selected, do you have the cash to get your movie into the proper format for a festival screening? Many events want at least a 16mm print with sync sound. This is not cheap to create, especially if you originally cut your film on video, sidestepping the whole film process as a money-saving measure. Sure, landing a screening at a big-time festival is quite a coup, but backing out because you can't meet the delivery requirements is possibly more damaging than never being asked. Remember, just because you screen at a big festival, you are not guaranteed a distribution deal with one of the major studios. Those who hit it big are literally one in a million. Are you willing to start the fund-raising process again to pay for a film print, when those funds might be better spent on promotion or even another film project?

How Will the Timing Affect Your Promotional Plan?
There's an understandable desire for every Hollywood wannabe to enter Sundance. It's almost like a rite of passage for independent filmmakers — make a movie, enter Sundance, get famous. The problem with that scenario is several steps are often overlooked. Sundance

happens once a year. And the organizers are very strict about the timing of entries. Let's say you finished production of your feature film in October and have a very strong feeling it is going to be the darling of the Utah-based festival. One problem: The entry deadline for the January event has passed. So do you sit on your film for a year until the next Sundance rolls around? Maybe you decide to enter other festivals in the interim. Think again. Sundance only wants to screen "premieres"; if your flick has been shown at the Podunk City Film Festival, Sundance (and an increasing number of other festivals) isn't interested. Thus, timing, or waiting, is also a cost associated with festivals. Is it worth it to keep quiet about your movie for a full year just to fulfill some festival's submission guidelines? Would that time have been better spent promoting and selling your movie or video? We entered *Pig* in Sundance. I even spoke with several Sundance people who, unbelievably, were familiar with the film. This response got our hopes up, and we followed some advice from a film-festival text about communicating with those people who watch the submitted films (known as **screeners**) and other staff to establish some sort of personal connection. We mailed individual promo packages with a postcard, sticker, and short note to every screener and staff member (information we gleaned from the Sundance Web site) and waited like everyone else for the official announcement. Like most everyone else, all we got was a rejection letter. To be honest, we put off promoting our film to both the media and buyers for almost six months because of our dreams of premiering at Sundance. Was it worth the wait? Nope.

How Worthwhile Is the Festival?

Obviously, big, media-frenzy festivals are opportune outlets for a self-distributor looking to gain attention for his project. And even the highbrow, yet less attention-getting affairs carry a certain amount of prestige that benefits those associated with the event. But not all festivals are created equal; some are merely moneymakers for their organizers, while others are less sophisticated than a grade-school talent show. At the least, check out a program from any festival you may be considering. This document speaks volumes as to the integrity and worthiness of the event. A photocopied-and-stapled flyer probably represents a festival you'd do well avoiding — unless, of course, your project fits that kind of genre (for example, an experimental, black-and-white, VHS-produced,

3-minute short film about your fingernail). Call the organizers and ask about sponsors (those companies like Kodak or Sony that may be contributing funds or prizes for the festival), prior years' attendance, special guests (will there be any agents, distributors, or buyers present?), and screening-room aesthetics. It doesn't do you any good to have your film "premiered" to nine people sitting on folding chairs in a rented strip-plaza office space (with the windows blackened). As excited as we were upon learning *Pig* had been accepted by the Ohio Independent Film Festival, we were even more disheartened when our sound and cameraman told of the horrible screening conditions of the now much-improved show (and even more so when we discovered that every Ohio-produced film or video was accepted). We should have taken the time to fully investigate the technology and organizational aspects of the event. Be wary of festivals with hefty entrance fees promising connections to "Hollywood insiders." That's often a scam that involves the organizers trotting out some has-been film or videomaker that has had no connection (or may never have had any association) with the industry for years and is merely making an appearance to collect a paycheck.

What Kind of Contacts and Exposure Will Result from Your Participation?
A well-attended and well-publicized event offers limitless opportunities for you to network with other film and videomakers, as well as buyers and the media. Even if you don't get an offer for your project — and many festivals are discouraging the "market" aspect of the event, trying again to focus on the art — the people you meet, the exposure your film receives, and the experience of showing your movie to a paying audience usually make up for all the expenses involved. Usually. Unfortunately, some festivals are populated only with those people screening a film or supporting someone screening a film. That is why it is so important to survey the organizers and their materials before submitting to any festivals. As cool as it sounds to say you were an "official selection" of the yearly Ypsilanti event, if nobody beyond other competing filmmakers views your project, you probably should have skipped it.

Is There a Cash, Product, or Distribution "Prize"?
An award of money is fairly common with many festivals, though some give away production assistance (in the form of film stock, cameras, or

editing equipment) and even distribution agreements for the winning entries. Video-themed festivals in particular are more apt to offer the product prize — quite an enticement to a producer looking for a low-cost way to acquire some gear. Any festival offering a distribution or sales connection should be carefully considered. Contact the organizers and ask very specific questions as to what's involved in such a set-up. You want to be sure that it's not just window dressing to play on the desires of hopeful film and videomakers. Confirm that the advertised connection is in fact a real connection. Sometimes, unfortunately, a festival will simply fly in some buyers and wine-and-dine them for a couple of days in order to promote the fact that acquisitions agents will be screening the winning entries. True, these folks may watch the films, but that doesn't mean they have, or ever had, any intention of "acquiring" one. It sounds pretty slimy, but it happens all the time (and not just in this industry — the same goes with music festivals preying on bands).

Has the Festival Launched Any Careers?
We're all familiar with the story of an unknown, independent, small-town filmmaker who made it big after his $4,978 feature — financed through the sale of his baby sister's hair — found rave reviews and created a bidding war among Hollywood's biggest studio buyers following its film festival screening. How often does this happen? That's a question you'll want to ask any festival organizers as you research their events. If the festival has been the springboard to bigger and better things for any attendees in the past, chances are good acquisitions executives, agents, and buyers will be in attendance looking for another hit.

Does the Festival Offer an Accessible and Attractive Location?
With more than 1,000 film festivals taking place annually, it's a safe bet that every one doesn't feature a locale boasting warm, sunny beaches, beautiful mountains, or cosmopolitan chic. Remember, the point of showing your project at a festival is exposure to viewers and potential buyers. If the event takes place in Hicksville, USA, 100 miles from the nearest jet-accessible airport, not only will nobody want to come, they couldn't even get there if they did. Though history and media attractiveness draw crowds, location is another aspect the self-distributor should investigate when surveying possible festivals. West coast events, even the smaller, less-known affairs, can draw industry players

more easily than those held in "middle America." Besides, if you are invited to attend, it's a nice bonus to travel to someplace you find appealing.

How Is the Market Aspect of the Festival?

As a self-distributor (and that is your first job regardless of the forum), the most important function of any film festival is its market or sales aspect — are you going to sell your project as a result of attending the event? All the awards and fans in the world are meaningless if you don't sell your film or video. True, some projects are designed strictly for the "festival circuit." The producers of short films enter and attend as many festivals as possible in an effort to raise awareness of themselves in the film or videomaking industry. These projects are meant to show off the abilities of the filmmaker, paving the way for future, bigger-budgeted productions. The self-distributor does not have that luxury. You are looking for a sale, an audience, a way to make money from the event. If a festival has no organized "market," you may want to reconsider your application. Sure, the publicity surrounding a good festival is worth all the costs. But how many festivals can be considered "good"? Look for festivals that openly promote the market side of the event, meaning they have opportunities to rent sales tables and set meetings with interested buyers.

Using film festivals as a promotional tool can be tricky. It can get expensive and produce no results. Or, you might attend one event and make a deal of a lifetime. Don't just enter a bunch of festivals looking only for acceptance. Believe me, with all the film and video festivals taking place at any one time, you will get accepted, no matter how unprofessional or inappropriate your project. There's something out there for everyone — but that doesn't mean you will benefit from it. You must approach festivals as a way to promote your film to a buying or viewing audience. Consider the investment versus the payoff. And don't always judge investment monetarily. Timing is a huge issue with festivals, and the amount you waste can often be worth more than any amount of money. If possible, talk to someone who's attended a festival you are considering, or attend it yourself if you can. Get the feel of the action, watch attendees and exhibitors, check out any industry dealings, and make it a point to speak personally with an organizer. Can the event really help you sell your film? That's the question you need to answer.

POSTERS, PUBLIC SPEAKING, POSITIONING, AND PUBLISHING

Guess what? After exhausting all the promotional techniques already described, there's still more left. They're not as sophisticated and organized as the methods already outlined, but the following promotional practices do have their place in a successful self-distribution plan:

Using Posters, Flyers, and Stickers

As old as the commerce of the entertainment industry itself, using posters to promote a movie or video is still an effective way to get the word out about certain projects, or to specific audiences. Let's say you've produced a video on the history of your small town. It includes interviews and comments from community leaders, historical photos and film footage, and beautiful shots of nearly every aspect of your village. Your obvious market is anyone living in or near the city, and what better way to reach this group than to hang flyers and posters advertising the video's availability. Many small shops, offices, libraries, and other public areas have bulletin boards for posting such things. Use your key art, include the price and a phone number, or better, a Web site where someone can obtain more information. This is a simple, free way to keep promotion of a project constantly in front of an audience. Check posting locations regularly to be sure flyers are still hanging and in good physical shape. It's also not a bad idea to constantly change the appearance and design of the piece. People get used to seeing the same flyer and will eventually ignore it. If you redesign it, however, they'll reread the same information thinking it is something new. Be careful where you hang any flyers, posters, or stickers. A local *Pig* promotion found logo stickers on hundreds of street signs and telephone poles around our county. A group of dedicated volunteers kept stacks of the large, 8" x 10" stickers in their cars, seizing opportunities whenever they could. We later learned the police weren't too fond of our defacing public property; luckily nobody got in trouble. Regardless, we got a *ton* of publicity and awareness form this guerrilla-level campaign. You literally couldn't drive on any road within a 10-mile radius of the city without seeing at least one *Pig* sticker. I heard about these "free" ads daily, and they only cost us about $25.

If you've printed mini-posters or stickers, ask area merchants if you can leave a stack on their counters. Video stores are an obvious choice, but also check with libraries, bookstores, entertainment retailers, and the like. It's human nature to pick up something — especially if it's free — so why not have your video or film promotion available wherever people stop to do business? You probably don't want to make these pieces too sales-oriented, but a small reference to a Web site or phone number is certainly fine.

Making "Celebrity" Appearances

Okay, so you're not yet listed in the Hollywood Creative Directory. But that doesn't mean you aren't a celebrity. How many people in your town can put the words "accomplished film/videomaker" on their resume? You've done something pretty outstanding and should be proud of the accomplishment. So proud, in fact, that you'll want to share your experiences with groups that like to listen to such stories. Many social groups, schools, and universities are always on the look-out for local artists willing to speak to their members and students. If there's a college or art academy nearby, give the public relations office a ring and ask about this possibility. As Kent State University (Kent, Ohio) graduates, my partner and I were asked to not only speak, but show *Killer Nerd* to campus groups about a year after the film was released. The movie is targeted at that age group, so it was a perfect match. The organizations you approach for this kind of publicity really depend on your project. A civil-rights documentary should easily land you in many minority-organization gatherings, while a video on fun and nutritional cooking will have you speaking to seniors' and women's social groups. The best part of this technique: You often get paid for your time, or, at the least, a photo of the event appears in the local paper. If you book one of these gigs, ask the officers of the group whether or not they pursue any publicity with local press. If not, take it into your own hands and write a press release detailing what is going to happen and when is the best time for a photographer to show up. You may be able to sell your product to members after your speaking appearance. Much like a book signing, you can set up a table offering to autograph special copies of your video or DVD to those making purchases.

Sometimes you can simply show up, sell tapes and DVDs, and sign autographs. Check with libraries, bookshops, and video stores in your area, selling yourself as a local celebrity. Ask to spend a couple of hours in their store doing your thing in exchange for free copies of your project. The store owner wins both ways — she gets a copy or two of your video to rent or sell and the publicity of staging a celebrity appearance in her shop. Don't forget to call the media and a bunch of friends.

Positioning Yourself in the Industry

While it's true that your main job as a self-distributor is to sell product, you also want to attain some level of recognition within the film and videomaking community as an accomplished professional. How do you do both? Speak at an industry-affiliated gathering. Many such opportunities present themselves during film festivals. In addition to showing films, those events always feature panel discussions with practicing filmmakers who may or may not be showing their work at the fest. Check festival itineraries and schedules for panel openings, and give the organizers a call if you think you have something special to offer. Your film may not be an award-winning candidate, but possibly you financed it in a creative manner, used technology in a new way, or secured distribution through unique channels. Any independent film or videomaking knowledge that you've accumulated or "invented" that would be enlightening to others in the field can prove for an interesting presentation. Cause-oriented festivals (those accepting only films that deal with a specific issue, such as the environment, gay rights, or Native Americans) also have openings for speakers to discuss the topics of their projects. So even if you are not an expert on filmmaking, you might be on the subject of your project, which makes you a viable panel member. Again, after the presentation, you can hawk copies of your product to those in attendance.

Like festivals, traveling genre conventions and expos offer the same opportunities. FrightVision is one such example. A horror movie "nostalgia, memorabilia, and collector's expo," the fair holds discussion panels in every town at which it stops, featuring local or regional independent horror filmmakers. These folks are paid for their appearance, have the opportunity to sell tapes and DVDs of their movies, get to sign autographs, and are presented as celebrities to those attending. Compile a mailing list of anyone you meet at these kinds of events. If they liked you enough to ask for an autograph or ask a question, they'll probably buy whatever you make next. At the least, the people on your list are good "word-of-mouth" publicizers for all you do.

You can also position yourself through club or organization member-ship. There are countless film and videomaking groups around the globe looking for members; check out what's available and join those that make sense. It's a good way to network with others in the field. Many of these associations distribute membership newsletters or mag-azines that are open to submissions. Talk to the editors of these pub-lications to learn about soliciting an article or photo for the next issue. Also check out organizations specific to your project's subject matter. Much of this research has been done in your search for sales outlets, and now is a good time to inquire about membership. Surely an envi-ronmental group would warmly welcome an independent videomaker cre-ating projects that support their cause, and a story in their association's newsletter (whether it is electronic or print-based) is a perfect way to spread the news. Your "recruitment" by the group looks good internally — officers can tout the fact that they are attracting active members. And such an article is a nice soft-sell approach for your product and expertise.

Using Newsletters to Promote and Stay in Touch
Though it involves a little more work than many of the other promotional methods outlined above, a newsletter is a great way to communicate news about your film or video to an audience. Whether it's delivered electronically via e-mail or through the traditional mail system, a two-page newsletter gives you the opportunity to report on project updates, cast and crew happenings, awards and achievements, and — most importantly — it gives readers the chance to buy a copy of your film or video. Layout can be done on any of the simple and inexpensive desk-top publishing programs available, with output and printing easily accomplished at a quick-copy center.

A series of newsletters designed to front-end promote *Pig* consisted of a two-sided, 8 1/2" x 11" sheet of white copy paper with black printing. After copying, the newsletter was tri-folded, stuffed in a business-sized envelope, computer-labeled and mailed with one first-class stamp. Each piece cost approximately 50 cents. If you create your newsletter electronically, it's virtually free. The downside to that method is that we only had about 30 e-mail addresses, whereas at the end of the cam-paign, we had a list of nearly 800 snail mail addresses.

Who receives the newsletter? To start, send one to all cast, crew, loca-tion owners, suppliers, family, and friends. Think about where you shot your project. Did you use a restaurant or a park? Who helped secure the location? For example, we shot *Pig* on a residential street that the local police closed for our use. To do that, we had to speak with the mayor, the city service director, the police chief, some patrol officers, several city maintenance crew workers (who set up the "road closed" signs), and each of the residents living along the stretch of road we used. Each one of those people received a copy of our first newsletter. As you consider your production and the number of people that made it possible, you'll be surprised at how quickly the names add up. Also add in any names you cull from your "celebrity" appearances and inquire about purchasing mailing lists from any associations you may join. At the least, if the group meets locally, you can hand-distribute the newsletters at any gatherings.

I've used newsletters in mailings to distributors, buyers, and agents during the self-distribution process of *Pig*. It's a quick way to give these folks information about the film and maintain visibility for the project. Video retailers and rentailers are another natural audience of readers, as are any potential buyers you may have discovered. As you make sales, be sure to include any current customers in future mailings. You don't want to simply sell your product, but cultivate a long-term customer. You want to maintain contact with these folks and turn them into continuing buyers of future films and videos and ancillary merchandise, as well as create a "fan" that will give you and your movies or tapes good word-of-mouth promotion. For less than half a buck, that's a pretty good deal.

A newsletter can be mailed on a regular interval (weekly, monthly) or whenever something special happens with the project. Be sure to include contact information in every issue and dedicate at least half of one page to selling your product and any ancillary items (t-shirts, hats, etc.) that may have been produced for marketing purposes.

Use as many photos as possible in the newsletter — production stills, behind-the-scenes shots, and publicity event coverage are all good ideas. If your movie or tape has received any media coverage, reprint those articles or reproduce the actual piece (with the source's permission, of course) within the newsletter. The *Pig* mailing list included more people living in other states and far-off cities than it did locals. Those people were not exposed to much of the press the movie received, so the newsletter provided a perfect forum for relaying that kind of coverage to a broader audience.

Chapter 21 Summary Points

- Publicity stunts are designed to attract the interest of both the public and the media in a unique way.

- Though they appear to be a spontaneous occurrence, publicity stunts are actually exhaustively planned events requiring strict attention to details.

- Working with local sponsors appropriate to a project allows self-distributors an affordable way to stage a premiere.

- Premieres can be viewed as the ultimate promotional project, and should be carefully orchestrated with regard to audiences, media, venues, and themes to guarantee a positive outcome.

- Projects screened at a film festival enjoy the benefit of exposure to audiences, potential buyers, media, and other film-makers.

- Costs, timing, location, and expectations of every film festival should be examined before entry.

- Speaking at conventions and other public venues is another promotional technique allowing you to position yourself as an "expert" in the industry.

- Regularly distributed newsletters containing timely and interesting information and photos are a low-cost and effective manner to promote your project to potential buyers and audiences.

PRICING

As your new and improved presentation package shapes up, and you prepare a list of the most promising buyers for your initial self-distribution efforts, one vital question still needs to be answered: How much are you going to charge for your film or video?

It's hard to put a dollar figure on a product that represents possibly two or more years of your life, but the time has come to do just that. Not only will you want (and need) this information on any sales marketing pieces (such as sell sheets, press releases, and some packaging), you also want to be secure in your pricing decisions when speaking with any buyers.

Depending on the market situation, pricing can be strictly determined by contract terms. Some companies have a set pricing structure, and you play by those rules. Rentrak, a leasing wholesaler, pays the same amount of money per tape or DVD, per supplier, per term. Other deals, however, rely on the tried-and-true method of negotiation. In this scenario, one of the parties (usually the buyer) offers a price. The self-distributor will try to improve on that initial offer, while the independent distributor, retailer, or other market buyer will obviously work toward the other direction. These arrangements are most common with foreign, broadcast, cable, and institutional sales.

An independent self-distributor's biggest potential lies with the home entertainment industry; sales to that market are somewhat different from the scenarios presented above. If a video store buyer is interested in a movie or tape, he'll want to know immediately if the price of the product is within his range of affordability. How does he determine that range? Well, market conditions are definitely a factor. For example, in the early part of 2001, DVDs had not yet saturated the market, and consumers were looking for as many new titles as possible in that format. Thus you had a sellers' market. If you had an independent feature that delivered on genre conventions and was available on DVD at that time, a buyer would be willing to go up on his spending.

Probably more importantly, though, a buyer determines his acceptable price range by considering his most profitable selling price point, and working down from there. As said a couple of times before, it's a business and everyone has to make a buck.

An independent self-distributor able to tap a niche target can often buck current pricing trends. Jimi Petulla's feature film, Reversal, is sold directly via his Web site for $29.95. Though this is nearly twice the going rate for Hollywood blockbusters released on DVD, Petulla has been able to sell more than 10,000 copies of the flick because of his ability to reach a very targeted market.

In the home entertainment market, prices paid to producers have dropped radically. Back in what was known as the "go-go days" of video (the early 1980s), the home entertainment market was prime for film and video product, creating an unquenchable demand that allowed sellers of all types of taped entertainment product to "set the price" in sales negotiations. Rental stores needed movies to rent and the studios, at that time, could only offer a limited number. The big boys weren't prepared to churn out movies as quickly as the thousands of camcorder-toting wannabes, so a $2500 feature made in an afternoon could actually find decent distribution. Unfortunately for independent feature film producers, that boom time has long gone. Too many shoddy films from novices with little production skill found their way onto video shelves. Though home entertainment is still a billion-dollar industry, inroads for self-distributed films and videos are not as open now.

Additionally, true independent producers (those working with budgets of $100,000 or less) are now competing with "independent" studio produced product. Movies boasting a $5 million (or more) budget are now deemed "independent" in comparison to the $100 million+ costs associated with most mainstream Hollywood projects. Thus, it becomes easy to see why truly independent self-distributed films and videos often face a challenging course in reaching the various home entertainment audiences.

With the advent of continually changing technology, wholesalers and independent distributors, as well as retail and rental video buyers, realize there is a limit as to how high they can price non-studio, independent product when selling to consumers and end users. Since it can be categorized as almost a "commodity" product, the lower the price of a video or DVD, the higher the sales. Home entertainment is not like an automobile. You can't really have "prestige" buyers that are more attracted to a product because of its higher price. A copy of *Monsters, Inc.* costs pretty much the same whether you purchase it online, at Target, or at Best Buy. However, if the pricing is too low, especially with an unproven independent film or video, the product may actually lose its appeal. How many times have you passed a bargain bin filled with "previously viewed" movies, most being titles you've never heard of before? The video store wants to unload unprofitable product, freeing up shelf space for more prestigious (and almost always higher-priced) tapes and DVDs.

A similar situation exists for self-distributors offering product to those outlets which will resell or rent it to an end-user. The goal is to price your film or video at a level that doesn't make it seem "cheap," and at the same time see some cash for all your hard work. How do you know what to charge for your product? That's not always an easy question. Surely, some very cheap movies have been sold for some very big figures. But that usually occurs in a formal distribution arrangement. At this stage in the game, it's basically a matter of haggling, trial-and-error, wishful thinking, and taking what you can get. There truly never is a "right" price, as you'll always think you should have asked for more after making a sale. Be happy with what you get, using the following guidelines to help you develop a fair price when selling to the home entertainment markets:

WHAT DOES THE MARKET BEAR?

When soliciting distributors and wholesalers, check out these compa-nies' marketing materials for product prices. Look to the trades. Read the articles about pricing, revenue, and economics of the industry. Examine distributor ads. What are they charging for a project of com-parable genre, length, and production value? Is it a suggested retail price or the actual price of the product?

WHAT IS THE DISTRIBUTOR'S REAL PRICE?

Trade magazines often advertise a product's retail price. But nobody pays that amount. Here's how it works: A distributor will buy product at a huge discount off retail — sometimes 60% or more. He then resells the product at a discount, maybe 20-40%, to resellers like Wal-Mart, Blockbuster, and others. These end-of-the-line retailers then sell at the advertised retail price, or maybe even a discount, too. In this chain of sales, there's room for everyone to make their "cut," and that's the only reason any of these businesses are still doing business. When selling a VHS copy of *Pig*, we had a suggested retail price of $49.95. As we started to market the title to wholesalers and subdis-tributors, the real price for them fell somewhere between $15 and $20.

When selling to video retail and rental shops, the discounting is even greater. In both single-outlet and chain-store sales situations, *Pig*, an 88-minute, shot-on-film, professionally boxed, and well-reviewed feature, was self-distributed for $20 a unit, two-packs for $35 — that's cutting it pretty thin. But it is what the market would bear. When a video store buyer has the option of bringing in an additional copy of *Gladiator* for $40 or paying the same price for a low-budget independent feature film that her customers have never heard of and may be a contender for the "worst-film-of-all-time" award, what do you think she's going to choose? Sometimes the price decision depends on eliminating the "no" factor.

HOW MUCH DO YOU NEED TO MAKE?

Making money is the point. Unless you view the process as a hobby, you probably want to at least breakeven, if not generate some profit. To do so, you'll need to basically divide your budget by a realistic sales goal in helping you determine an appropriate price. Depending on the genre, an independent, self-distributed movie or video can sell anywhere from 50 to 25,000 copies. That's a big range. Obviously, if you spent $50,000 making an instructional video on light-bulb mainte- nance, your audience is very limited, and you'll never breakeven unless you can find 500 generous people to shell out $100 per copy. But if you make a fairly competent shot-on-digital horror movie for $25,000 and package it attractively, your potential viewership grows substan- tially, bringing the breakeven price down. It is not unheard of to sell 2,000 copies of a genre product (remember, there are more than 15,000 video stores out there); at $20 a copy, you'll more than cover your production budget.

WHAT IS YOUR GOAL?

Go back and examine your goals. Do you want fame or fortune? An income or an image? If you are merely looking for recognition, exposure, media attention, and placing your tape or video in front of the viewing public, then attracting buyers with a very low price point that just covers your costs may be the right choice for you. On the other hand, those filmmakers with investors breathing down their backs must investi- gate pricing levels that will return the highest profits without scaring customers away.

Regardless of your motivation, setting the right sales price for your proj- ect is a very important decision. If, after a couple of unsuccessful solic- itations, you get the feeling that buyers are passing based on cost, ask the next prospect what he feels is a fair price and adjust accordingly. You don't want to cover too much of the market only to find out that a couple of dollars were keeping you from widespread sales. Remember, a "no" is a "no"; buyers will not react positively to a second round of lower-priced product. They'll feel like you were trying to cheat them the first time. Besides, how do you justify such action to those buyers who did purchase your tape or DVD at the original higher cost?

Chapter 22 Summary Points

- It can be hard to determine a dollar figure for the worth of a film or video project that represents two or more years of your life.

- At times, the price you are paid for your "product" is predetermined by a contract, but more often than not, it comes down to negotiating terms.

- Self-distributing to the home video market means setting prices based on current industry trends and the buyer's financial situation.

- Since it competes with Hollywood "A" product, an independently produced film must be priced competitively to sell, though not so cheap as to suggest inferior quality.

- Before setting any prices, consider your ultimate goal and what the market will bear, then determine a buyer's discount.

THE MAILING

Sooner or later, whether you feel ready or not, it's time to start the solicitations. The first step in this process is the mailing: sending your screeners, sell sheets, cover letters, and other assorted presentation-package items to the potential buyers you have identified. There's really no mystery to this step — you basically want to get your solicitation materials in front of the people who will decide whether to stock, broadcast, program, rent, or resell your film or video. Some things to keep in mind when commencing your mailing campaign:

Keep It Workable

At this point, you may have hundreds of potential buyers' names and addresses. You don't want to send materials to this whole group at once. Do your mailing in groups of 10, restricted to similar outlets, following the chain of demand. For example, if your project doesn't meet the needs of PPV providers, you'll want to start with the home entertainment industry, sending information to 10 local single-store video rental facilities. In this way, when you start callbacks, you won't have to worry about mentally juggling market-needs information. You will be talking to virtually the same person at each location. A local single-store buyer or owner has the same needs as any other single-store buyer in his geographic business area. As your sales skills develop, and you become more confident, you may want to solicit buyers in different industries concurrently. But at first, make sure you are not taking on more than you can handle, because it is too easy to become discouraged.

Send a Teaser First

If possible, pique interest in your project by sending a "teaser" postcard, P.O.P. item, or letter before the main package. This is especially helpful for projects that have received little-to-no media coverage (or at least no coverage that a buyer would have seen). A postcard touting the *Pig* Web site, and little else, was sent to potential buyers approximately two weeks before a solicitation to buy. In that way, recipients could access information about the film via the Internet, if so interested, or at the least were aware of the film and its title when the sales materials landed on their desk. Promoting your project in some way to buyers before you actually solicit them with marketing materials takes some of the sting out of a blind mailing.

Keep Letters Short

This is a no-brainer. Buyers are looking for pictures and prices. Look to the sales-letter instruction presented earlier and heed that advice. Point to a good review or interesting hook and ask for the sale. Tailor your letters to the market approached. If hitting local buyers, aim your plea at a local angle (project made in the town, support local artists, community interest). When soliciting wholesalers and subdistributors, talk of affordable pricing, good packaging, and guaranteed delivery.

Handwrite the Address

How much mail do you simply toss because it's addressed with a computer-generated label? Hand-written screams to a recipient that there's something personal inside the envelope, that the materials are only being sent to him and nobody else. Sure, it takes extra time, but it truly increases interest and readership.

Consider the Buyer When Stuffing the Envelope

A video retailer mailing can consist of two sheets of paper: the sell sheet and a cover letter. Don't send 13 pages of reviews and press clippings. The retailer won't read it, so your sales message will get lost in the trash. When soliciting a potential broadcaster, you probably do want to include reviews and media clippings, as well as a screener. They don't buy 25 tapes or DVDs a week, so they need more information to make a decision. Make it easy for a recipient to know the purpose of the mailing — you want to sell them your movie or video.

Forget about Order Forms

Video buyers do business by phone or through the Internet, not by mail. Our initial solicitation to retailers for *Pig* included an order form and return envelope. Out of the 100 mailed, none were returned. Many of these outlets purchased the tape, but not by mail. Don't waste extra printing and mailing costs with these unnecessary materials.

Dress Up the Envelope

Did you print stickers to promote your project? Place one on every envelope you mail to prospective sales outlets. Again, it gets the name and any images associated with your project in front of the buyer before they read your materials. Anyone handling the package will also be exposed to this simple promotional technique.

Use First-Class Postage

Don't ship mailings any other way. Bulk rate, 3rd class, and media-class shipping can take days to deliver and is handled with less care.

Chapter 23 Summary Points

- Don't let enthusiasm get out of hand during your first mailing — keep it simple and organized

- Mimic the studios' practice of whetting an audience's appetite for things to come — send a teaser before the formal solicitation.

- Consider the recipient of any solicitation and only include those items that are necessary to make the sale.

CALLBACKS AND
MAKING THE SALE

Most independent film and video producers make fine self-distributors and have no problems selling their project. They are most intimate with the production and know best how to promote it through a sales call. Selling, however, can be the hardest part of the process for others. Some take-to-the-heart advice: If you are uncomfortable asking people to buy your film or video, find someone else to do it — real quick. You don't have to be a polished salesman to make sales, but you do have to possess some charisma and even some "slickness" if you want to see results from all your efforts up to this point. As a self-distributor, the buck stops with you. If you don't or can't sell your movie or video, sales are not going to happen. There will be no knight in shining armor riding up to your door with distribution offers. Any reluctance and pro-crastination will affect your success. You have to be willing and able to call buyers, ask for a sale, and accept rejection. If any part of that process is unappealing, look to colleagues and others involved in the project who may be better suited to handle this step.

Regardless of who makes the call, give any potential buyers at least 4-8 days (depending on location) after a mailing before contact. You want to give them enough time to receive and open the package, but not too much time to forget about the contents. Self-distribution differs from your initial solicitation efforts — you are now primarily responsible for contact. If a studio distributor liked your project, chances are good they would call you. You were looking for representation at that point. Now you are representing your project, functioning as the supplier. Make no calls, get no sales. It's that simple.

Ask for whomever the materials were addressed to, and when speaking with this person, immediately identify yourself and inquire as to the receipt, and perusal, of the package. Depending on the type and size of the company you are soliciting, this buyer may or may not have ultimate purchasing authority. In the home entertainment environment, buyers are usually able to make this decision. When it comes to PPV, cable, broadcast, and some retail businesses, the buyer sometimes only makes recommendations to purchase (recommendations which are usually followed).

There are countless books on selling techniques; everyone has his own style. You want to develop some sort of rapport with the person on the other end of the phone, and not just blurt out, "Do you wanna buy my tape?" five seconds after she answers the phone. You don't want to talk too much, either. After a couple of calls, you'll know when the time is right to ask for the sale. Don't dance around the issue and hope the buyer will bring it up. Plainly ask if she is interested in buying your project for whatever capacity her outlet represents. If you get a negative response, try to elicit further information on why she said "no." Always thank her for her time and ask if you can continue to solicit her with new product in the future. If she'll never be interested in buying independent, self-distributed films and videos, you don't want to waste any more time and money sending her materials.

If you hear that magic "yes," quickly thank her and confirm the following (that may or may not apply to all buyers):

- Shipping/delivery address
- Invoicing address and contact
- Quantity ordered
- Total price of invoice
- Expected delivery date
- Method of payment
- Shipping method

By the way, feel free to celebrate after a successful sales call. It's been a long trip to this point, and the fact that someone has agreed to buy something you created is validation of your skills as both an artist and promoter.

METHOD OF PAYMENT
This really depends on the type of outlet to which you are selling, any contract terms that might be in effect, and personal/corporate preference.

Payment from sales to wholesalers, large independent distributors, broadcast, cable, and PPV outlets will be dictated by the terms set forth in a contract. What is customary? Nothing, really, except don't expect payment for up to 30 days or more (usually more). Contracts are a wily device, and are covered in detail in Chapter 25.

For sales not bound by a contract, the method of payment can vary. Most small-to-mid-size video retailers and rentailers will prefer to deal on terms or "cash on delivery," commonly known as C.O.D. **Terms** means an invoice is issued, but not due for payment for a period of time, or a "term." Offering 30-, 60-, and 90-day terms are the most popular increments used. *Pig* was sold to most video retailers on 30-day terms. How did that work? I would contact a potential buyer and after an order was verified, the tapes were shipped to the specified address. Under separate cover, and sometimes to a different address, an invoice was shipped on the same day as the tapes, indicating a total amount due in 30 days. We never included the invoice in a product shipment because in many instances, the person opening packages is not the person processing invoices.

Only on several occasions did someone request a C.O.D. shipment. This method of payment actually benefits the self-distributor because C.O.D. service allows the mailer to collect payment for merchandise when it is delivered. C.O.D. service can be used for anything mailed first-class, registered, express, priority, or standard. The merchandise must have been ordered by the addressee, who has the choice of paying at the time of delivery by cash or personal check. There is a small fee charged for the service; the process can take place at your doorstep or in a post office.

Accepting credit-card purchases is probably not worth the expense of arranging until you start selling to individuals and through any online applications you might create. Most video retailers, distributors, and other organizations will pay with a company check. It's unlikely you'll be asked to process a credit-card transaction, though some very small retailers may request the service. Once you begin doing business on the Internet, a host of credit and electronic payment applications are available at very reasonable prices. Accepting these kind of convenience payment methods encourages impulse purchasing and drives most Web-based commerce. You will definitely want to take advantage of this device when your self-distribution efforts turn toward consumer and other Web-reachable groups.

RISK

At this point, you may be thinking it sounds pretty risky to ship tapes or DVDs to someone you've never personally met, in hopes that they'll send you a check sometime in the near future. And you know what? You're right. This risk is one of the main roadblocks to a successful self-distribution effort. Unlike the deep-pocketed studios and multi-million-dollar independent distributors selling a continuous line of product to every market imaginable, the self-distributor has one film or video and no financial resources to float credit. If you get ripped off, you don't have another project already in the sales pipeline or 50 other buyers banging down your doors to buy more. A single bad debt might sink your operation.

A subdistributor in Memphis, Tennessee, ordered 50 copies of *Killer Nerd* during the height of the film's sales. We were a little hesitant to fill the order since this gentleman's company was unknown in the industry, but he told me he specialized in reselling videos to small rental stores throughout the southern part of his state. Enthusiasm to get the film in as many national locations as possible won out, and we shipped the tapes. About two weeks later, a check arrived and our worries were eased. Three days later, the subdistributor called again, this time ordering 100 tapes. You can imagine our excitement! This order represented more than $2500 in sales, not bad for a film that cost less than $12,000 to produce. Since this buyer had "good" credit with us, we duplicated, packaged, and shipped 200 copies of *Killer Nerd*, and never heard from the subdistributor again.

Most self-distributors have similar war stories to tell. How do you protect yourself against such a loss? That process can be tricky. Obviously, you could run credit checks on all potential buyers. But that's a lengthy and expensive process that can ruin impulse buys and turn off a lot of folks as well. You could also allow for C.O.D.-only orders to guarantee payment. Again, not everyone is prepared, or wants to do business in this manner. It comes down to weighing the value of a potential sale against the risk of being ripped off. Local and regional retailers are in physical striking distance so you can always contact them in person if a sale goes sour. Recognized wholesalers, large independent distributors, broadcast, and cable outlets are safe bets as well — they didn't get to their positions in the industry by not paying their bills. The real danger comes when selling to far-off single-outlet video stores and small-time distributors. Unfortunately, a number of these businesses make a habit of preying on self-distributors and other vendors unable to financially combat dishonest business practices. Basically, they know they can steal product from you without recourse. Sure, you can badmouth the company and refuse to send them any more tapes or DVDs. But that doesn't hurt these players. Bottom line: They're profiting from your loss. They've been doing it to others before you and they'll do it to film and videomakers of the future.

If you have the time, ask a new buyer for a list of other vendors he's dealt with and make a few quick calls to check the company's credit worthiness. Any reputable business should have no problem with this tactic; if they do, that's a sure warning sign that something may be amiss. While this tactic may help weed out the good from the bad distributors, it's not a feasible tactic to use with those retailers ordering one or two units of product. With such small ticket sales, the time and expense involved in tracking a buyer's references is not worth your energy. You need to be selling as often as possible, and any time taken from that activity lessens your chances for increased profits. To be honest, it's monotonous and mentally draining to spend half of your time checking credit records. It really takes the "fun" out of the business of being a self-distributor. For these kinds of sales, the best advice is: Trust your gut. Not very scientific, I know. But it's really all you can go on if you want to work at selling your film or video to as wide an audience as possible. Try to have faith in humans, too. Most people you deal with will pay their bills. As for the others, chalk it up to paying your dues. Even when you don't get paid, at least your film or video is finding viewers, and that's always worth something.

Chapter 24 Summary Points

- If you don't have the confidence to make a sales call, find someone who does.

- In the self-distribution scenario, all responsibility for contact calls falls to the seller.

- Callbacks must be timely, courteous, and professional — you'll be competing with every other product supplier (including the studios).

- Unless described by a sales contract, the buyer's method and timing of payment to a self-distributor can vary widely.

- The risk of incurring bad debts is one of the costs of operating as a self-distributor.

CONTRACTS

Regardless if you're signing with a traditional studio distributor, large independent, or self-distributing your film or video, a contract will most likely be involved at some step of the process. Low-quantity sales to individuals, rental shops, and retail stores won't require this kind of paperwork. But when you arrange a long-term sales agreement that finds some other company marketing your project to their customers on a continuing basis, a contract outlining the how's and when's of the product and money exchange is necessary. This legal document attempts to protect all parties from any unfair business practices, harm, and damage, and is a necessity. The uninitiated may jump at the chance to sign a distributor's "standard" contract. But this so-called standard or "regular" distribution contract is beneficial only to him. Business is business, and it is the buyer's job (regardless of how nice he may appear) to get the most he can out of you. However, don't worry about being muscled into a contract that seems unfair or makes unreasonable (which will inevitably lead to unprofitable) demands. If you are properly prepared and have a firm disposition, you can make counter demands to amend any contract offered you, with the eventual creation of an agreement that is acceptable to everyone.

Before entering into any contract negotiations, you must determine if the prospective buyer represents a reputable firm. Check the company's history. Though many honorable distributors open shop every year, and there always has to be a "first" client, try to find out how long the business has been in operation. Check the Better Business Bureau and Chamber of Commerce in the buyer's city. Is the company on record? Have there been any complaints? Do they even exist? I have shipped tapes to a subdistributor only to find he has gone out of business (actually, he just began operating under a new business name) 30 days after my invoice was due. These guys operate by convincing producers to give them credit and ship an order of tapes or DVDs. They sell the product, making 100% profit, close the business, and reopen again quickly as a new entity. The situation is very dangerous for first-time self-distributors pumping a lot of personal funds into the marketing effort.

Another way to gain insight on a potential distributor is to talk to other distributors, retailers, video buyers — even the media — to inquire about the distributor's behavior within the industry. Is the buyer an accredited member of any entertainment-industry organization? How does the company handle its own clients? Is the company on good terms with other dealers?

With the business background check complete, it's time to start pre-contract discussions. Terms in a contract are wide and varied, and some truly are standard. But investigate the following closely, being sure you fully understand the implications of each, and what it means for both your film's future and any possible revenue you will get from the deal:

Exclusivity

Every distributor you deal with will want to be the only seller offering your film into the market. It's understandable; they'll be spending money to promote your film to buyers, so why should any other company cash in on this publicity by selling the film as well? There is some truth in that situation, but only if the buyer is indeed enthusiastically promoting your film as promised. It really depends on whether you are being courted to enter into a traditional distribution agreement or you are self-distributing. (This point is closely tied to "rights" — see below.)

If traditional distribution, check the buyer's sales area before granting any exclusive rights. Does their marketing reach include enough regions to adequately sell your film or video to all markets available? For smaller distributors, the answer to this question is usually no, meaning they depend on a system of subdistributors and resellers to get product to every niche imaginable. If this is the case, ask for a list of the subdistributors so that you can check on their business practices as well. You may feel that having your product passed between too many sellers could dilute the impact of any publicity and marketing efforts enacted to this point — those at the end of the chain might not be as excited about the project, expending little energy to sell it. Unfortunately, there's not much you can do about that; just be glad you landed a traditional deal that will get your product to a wide audience.

Those self-distributing have the option of allowing a mainline distributor to pass product through the chain, or dealing with any subdistributors directly. In essence, you will not grant exclusivity to anyone. You can't afford to when distributing on your own. For example, let's say a small distributor likes your presentation package and wants to represent your feature film. Though he boasts of a large marketing reach, you find he really only sells to the domestic video market — more specifically, he markets VHS copies to video rental stores in the southwestern United States. If you grant this business exclusivity to your product, you'll tie up all your foreign, cable, Internet, institutional, and other media (DVD) opportunities, not to mention video sales in the rest of the country.

Self-distributors should only discuss exclusivity with resellers who can adequately represent a segment of the market (such as a foreign-sales agent or Internet-streaming buyer). These firms and individuals specialize in a niche, and, more likely than not, can penetrate that aspect of the market better than you can on your own. In other words, you may sign several different contracts, all guaranteeing exclusivity toward a different segment of the entire market. This means you would work only with that specific buyer in placing your film or video into that portion of the market. The bigger independent distributors are not keen on this kind of arrangement, and will want you to grant them exclusivity over all markets. But if they believe in your product (believe it will make them money, that is) and you stand strong on this important point, you should be able to negotiate what you want.

License

Again, this contract point is very dependent upon whether you are working in a traditional arrangement or as a self-distributor. Those film and videomakers being courted by a studio or major distributor may be asked to sell their "ownership rights" to the film. This means that you are selling ownership of the product to the buying company. Selling your film to Paramount, MGM, or Warners probably isn't a bad idea. The connections and money involved in such a deal should keep you busy in the entertainment industry for the rest of your life. Selling your film to a small independent distributor is another story. In this case, you'll grant the distributor a license to sell your film. This means you are giving the buyer the right to sell your product in the film/video market. You are not selling your film to the buyer, you are just licensing him to sell it.

Rights
These are the areas and markets available in which to market your film. Every distribution company you speak with will initially want all rights to your film or video — they want to be able to sell your product on video, DVD, cable, PPV, broadcast, theatrical, foreign, and domestic. Total rights can also include merchandising rights — stuff like toys, games, clothing, posters, and more. Do you really want to tie up all markets available with only one distributor? Sure, if it means your product is going to appear in every video retail and rental store across the country. However, if a potential buyer cannot show you concrete evidence of a major marketing operation already in place, I'd suggest against it. The possibility of one distributor successfully promoting and selling your low-budget independent film or video to all available markets is pretty slim.

As a self-distributor, you will want to splinter rights as much as possible. Though distributors will want as many rights as possible, when you are functioning as your own distributor, you need to get your hands into as many "pockets" as possible — it's the only way you can be sure your product is being marketed to its fullest. Upon initial execution of our distribution agreement, Hollywood Home Entertainment wanted all rights to *Killer Nerd*. To be honest, we didn't know a whole lot about the business at that point, but a local attorney (with no entertainment-industry background) did question the section of the contract concerning rights, and advised us to limit the period that clause would be in effect. Luckily, we listened, as Hollywood Home Entertainment's strength was strictly in the home video market — they didn't sell foreign, cable, institutional, or much of anything else. After our six-month deal of granting them all rights ended, we took the film to different buyers (foreign, leasing wholesalers) on our own. Hollywood Home Entertainment still sold to the domestic video market while we worked with other parties to penetrate more of the total market.

Advance
This term describes any up front money given to you by the buyer, which will be subtracted from future earnings. Though an extremely rare practice in

the independent film and video market today, some distributors still offer an advance if they are very positive about the performance possibilities of a project. It is basically "good faith" money, securing your agreement with the buyer. Advances are only given in exclusive, all-rights agreements, so the self-distributor will never see this item on a contract. When granted an advance, you earn nothing on the sale of your film or video until the amount of the advance has been earned back through the formula set forth by the distribution company. For example, if you receive a $50,000 advance on your feature, and you are to earn 5% of net profits, the film must gross $1,000,000 plus the amount of any expenses (advertising, travel, prints, packaging) it cost to earn that million before you start earning any more money from the sale of your film. This example illustrates the fact that in most cases, an advance is the only money a producer will earn from the sale of his film. It is too easy for a distribution company to "hide" earnings in the expense ledger.

Time Frame

All contracts are very specific as to the amount of time the agreement between the two parties is valid. Self-distributors will want to negotiate an open-ended time frame, which allows the distributor to sell your project for as long as they like. This is a good arrangement in non-exclusive contract situations where you may be dealing with multiple sellers. Each party has a basically limitless amount of time to expend all of their marketing resources in promoting and selling your film. Small, niche sellers may need more time to properly move your film to their buyers than a large, studio-funded effort, thus the attractiveness of an open-ended agreement. But if you are in an exclusive situation, negotiate as short a time frame as possible (3-, 5- and 7-year contract lengths are common, though you might push for one or two years at most). If an exclusive buyer is experiencing slow sales of your project, they may stop promoting it. Your film or video will be tied up, without a sales effort, until the contract expires. Talk about frustrating! This is the main reason for obtaining an advance in an exclusive marketing situation.

Pricing

When negotiating sales of your film or video to the video retail and rental markets, try to obtain a guaranteed price per tape instead of a percentage of the sales price or total sales. If you demand $15 per copy sold, then it doesn't matter if your distributor markets the project at $59.95 or $29.95 — you will still receive the same amount per unit sold. If you agree to a percentage of the retail price, say 20%, you would earn $12 or $6 according to the retail prices given above. At $6 per unit, you may not even be able to cover costs incurred from making the project. Distributors will always try to convince you to take the percentage deal (especially a percentage of net sales) since it only benefits them. These arrangements allow for the inclusion of a variety of expenses that only serve to lessen your earnings.

Expenses

This determines what items the distributor will pay for and what items you, the producer, will cover. Advertising, point-of-purchase items, sell sheets, shipping, and packaging are just some of the items that will be included in this category. Distributors are notorious for cutting into your profits by piling on the expenses they supposedly incurred from the sale of your film. An outrageous personal experience: Our distributor deducted the cost of mailing a royalty check to me! The check was sent via priority mail, and on the attached accounting record, a postage deduction of $2.90 was noted for that very package. If you are providing finished product (labeled, packaged, and shrink-wrapped copies of your project, ready to sell), you have more leverage in making the buyer pay for expenses. Buyers should always pay freight charges for product going to their distribution centers. When sending an order to a distributor, always insist on shipping "freight collect," otherwise you'll end up footing the bill. Try to think of all the possible expenses that will arise in the sale and distribution of your product and a reason why you shouldn't be responsible for each cost. You won't always be successful, but it helps to be prepared for the worst.

Advertising and Promotion

One expense that you'll want to discuss in detail is advertising and promotion. You need to be sure any distributor handling your film or video is willing to back up their sales efforts with some sort of promotion to the markets solicited. Many small and independent distributors rely solely on their call rooms and established accounts in marketing new titles. It's less expensive than creating and buying print ads or direct mailing pieces to simply return to past buyers with a new product solicitation. If their buyers are local and trust the distributor to represent quality, attractive product, then sales are fairly predictable when compared to recent projects. This still doesn't guarantee you or your project much exposure in the industry. To the distributor's buyer, you are just another product coming down the pipe. There's no "buzz" or excitement about the film or video. If an ad campaign or promotional event precedes the solicitation, however, it makes your project seem like something special, and chances are much better you will attract a larger group of buyers or a larger order from those buyers making a purchase. This is the whole theory behind a limited theatrical release. The publicity afforded such an event spills over to audiences when the product enters the video/DVD markets, increasing awareness and sales. At the least, you'll want any distributor to place an ad in the video trade publications (*Video Store* and *Video Business*) to promote your film or video to the biggest potential market. Self-distributors may place such an ad on their own, listing all resellers handling the product. Direct sales pieces, such as the sell sheet, are another popular way films and videos are promoted to the home entertainment market. Who pays for the creation and distribution of these kinds of promotional techniques are needs to be addressed in the contract.

Recoupment

If the buyer is handling a sizeable amount of the expenses, they may ask for recoupment of all or a portion of these costs from sales realized. In other words, they will still be paid for the costs of advertising, promotion, travel, shipping, and other expenses — only not up front. Instead, these monies will be "recouped" from the sale of your project. The distributor still gets the money you tried to keep away from him when haggling over the expense section of the contract! This may not seem fair, but it is the cost of doing business in the entertainment industry. If working with a reputable and well-connected distributor, his links to markets and sales will be well worth the cost. But the inverse is also true for those self-distributing. You'll want to limit recoupment as much as possible if you've already done 90% of the job selling the product through your promotional efforts. Also, be wary of what portion of earnings the recoupment is subtracted from. Insist recoupment comes from gross sales, not net sales. Gross sales are total sales of your film or video without any expenses (sales salaries, mailing costs, advertising, etc.) deducted. Recoupment out of net sales only allows the distributor to dig deeper into your pockets.

Producer Input

When signing with a mainline distributor, there's a good chance any art-work or marketing materials or messages you developed — possibly even the title of the film or video itself — will be altered or completely discarded. As the "parent" of the project, your input with regard to artwork and promotional materials used in advertising the product is usually regarded, but not always followed. Studio and large independent distributors have been selling movies and videos for years, so they know what works with any given product in any given market. That doesn't mean, however, that your ideas are not viable. How important is it for you to have input in these decisions? To some producers, it's worth killing a deal, while others are just glad to see their hard work finally reach an audience, regardless if the marketing message and look of the promotional materials in no way represent the original feeling of the piece.

Also to consider: What rights do you want to grant the distributor/buyer to edit the project for purposes other than meeting censorship require-ments of broadcast situations and other market restrictions? When we were shopping *Pig*, one distributor interested in the project wanted to add a musical score and add some scenes of violence. The whole point of the movie is to mirror the often bland existence of a small-town police officer. We did not want the movie filled with music which detract-ed from the quiet and unsettling ambience of the patrol cruiser and backroads where much of the movie takes place. Frankly, I wanted a traditional distribution deal so badly, I would have changed the movie to a comedy if asked. But, in the end, there was no way we were going to add music to the film. Was that a mistake? Should we have taken the deal and compromised our artistic intent? Maybe, if it were Miramax or Fine Line Features dangling that kind of deal in front of our noses. The distributor in question wasn't of that caliber, so — after weighing the pros and cons of changing the movie vs. the potential increase in sales — we decided to stick to our filmmaking ideals. Ask me again today, and I might have a different answer.

Credit

Many buyers, including distributors, will want you to extend credit to them when shipping product. Here's how it works: A distributor will order a quantity of tapes or DVDs from you to sell to their clients. They'll extend at least 30-day terms to their customers, and want the same from you, meaning you won't receive payment for your product until at least two months after you've shipped it out the door. Do you have the cash flow available to float this kind of credit? Can you trust the distributor and/or buyer with which you are about to do business? That is why it is so important to investigate the business practices and background of any company you enter into a sales agreement with. Be sure to make it clear in the contract the absolute maximum number of days allowable before payment is due. Terms of net 60 and quarterly payments are common with many distributors. If you feel uneasy about shipping product without advance payment — and C.O.D. is not an option — limit the number of tapes or DVDs you send to a single distributor until you are assured of their regular paying habits. Getting stiffed for a shipment worth $100 is a lot easier to deal with than getting stiffed for $1000.

Right to Inspect and Audit

This allows you the right to have a third party (certified public accountant or mediator) inspect and audit the buyer's books if you believe there to be a discrepancy in the sales of your project. Be warned that this is not a fast and cheap option — accountants cost thousands of dollars and it takes some legal wrangling (not to mention lawyer's fees) to even arrange this kind of action. You do want it included in the contract for safety's sake; it can come in handy if you find yourself with a run-away hit on your hands. Probably the most famous example of an independent filmmaker never seeing his rightful earnings from a low-budget movie that went on to become a national phenomenon is George Romero and his *Night of the Living Dead*. Though the film grossed hundreds of millions of dollars since its release in 1970, Romero and the other producers saw little of that money due to some creative contracting and bookkeeping on the part of their original distributor. Worse yet, that legal hassle allowed the film's copyright to lapse, leaving it fair game for public-domain use. Thus the dozens of versions now available from a number of sellers on both video and DVD. The film also appears from time to time on broadcast and cable television, especially around Halloween. This extreme case points to the importance of carefully reading and fighting for contract points that will protect you and your property.

Right to Terminate

If you are unhappy with the distributor's handling of your film or video, you have the right to terminate the contract and seize or have returned all product remaining at his place of business. This is important for film and videomakers with "trendy" or "flavor-of-the-moment" type projects. Let's say you created a reality video that deals with hidden-camera dating encounters. Excited by a current wave of reality entertainment popularity, a distributor orders 500 DVDs of your program, positive they can sell the DVDs to his customers. Well, the reality fad passes, and your distributor, who ordered high quantities of reality programming from other producers as well, is now telling you they can't pay their bills on time. You don't want your program tied up with a distributor who isn't promoting it, nor do you want them to possess 495 copies on what appears to be an endless credit period. The right-to-terminate clause allows you to end your dealings with them due to such a circumstance, forcing them to return all of the product – either on their own or by force of the law. Sure, it's depressing that you didn't make any money from the order, but at least the DVDs are back in your possession.

Deliverables

The most problematic portion of a contract for many independent film and videomakers is the part describing your responsibilities with regard to deliverables. Deliverables, commonly stated in contractual language as "delivery of items," describes what you need to give the distributor or buyer to make the contract effective. This is your end of the deal. It's more of a burden for those entering into studio-backed arrangements, but even a supposedly simple, self-distributed, home-video deal can find you searching old files and spending some money. In order from cheap-and-easy to costly-and-time-consuming, be prepared to ante up any or all of the following:

- Full list of the film/video's credits – Be sure to include all cast, crew, special technicians, labs, location, and product providers and anyone/any place else that contributed to the completion of the project.

- Copy of the script – Some distributors may want the shooting script, some will want the production script, others will ask for both. Have clean, bound copies ready.
- Cast and crew biographies – A detailed description of all past credits, education, and training of all principle cast and crew involved in making the film or video.
- Dialogue sheets – Necessary for foreign sales, this is not a copy of your project's script, but rather a word-for-word transcription of all the spoken language (dialogue and voice-over) as it appears in the final version of your film or video. You can compose these sheets yourself (probably by memory if you've watched your film as many times as I've watched mine) by simply writing down everything heard, or through a service which charges anywhere from $250-$500.
- Music cue sheet – This can be very simple or very complicated, depending on the project and the amount of music used. Music cue sheets are a listing of each music cue used throughout the project, with accompanying song titles, duration of the cue in minutes (in tenths of seconds), artists performing the music, and publishers (BMI, ASCAP). Music cue sheets are necessary for television sales, as broadcasters and cablecasters maintain licensing deals with the music publishing companies to pay royalty use rights. Hopefully, you didn't finish your project using unauthorized copyrighted music. You may think that four-second guitar effect is unrecognizable, but when it comes time to signing a contract, are you willing to say that all the music in your project has been properly cleared? The soundtrack in *Killer Nerd* was a combination of scored music and original regional band recordings. All bands signed a contract granting us full use of the music, and the publishing company, if any, was contacted with the same information. Since the music was original, the bands owned all the rights, making music cue sheet creation and clearance very simple. *Pig* features only one original composition that lasts approximately two minutes. Our music cue sheet is one line long.
- Trailer – Though not all companies will ask for a trailer, since many will cut their own to meet their specific marketing needs, hopefully you've created one for promotional purposes before

this point. If not, carefully script one on paper to save time when going back into the editing bay. You should be able to cut a three-minute or less trailer in one day's studio time.

■ Publicity stills – Discussed previously in this text, it is vital to your distribution agreement to have plentiful and good quality color and black-and-white still photos from the production of your film or video. Distributors and buyers will use these photographs in the design of all marketing materials and count on them as the first encounter most consumers will have with your product. For many instructional-themed videos, photos can be easily staged after the fact. That's not the case with a feature film where large casts, crews, and sets may be involved. To fulfill many delivery terms, be sure to have good photos of every scene, staged shots for publicity purposes, head and full-body shots of every main actor, and behind-the scenes photos of both cast and crew. Director and director of photography shots are the most used of this kind.

■ Poster, one-sheet, postcards, and other marketing materials – This is very dependent on the kind of deal you are entering. Those signing with studios and large independent distributors won't need these items. Small, direct-to-video, and English-speaking foreign buyers may stipulate such materials as necessary to the agreement. Self-distributors should be well prepared with these items as they are essential to their marketing and promotional efforts. Producers counting on studio deals still may want to design and produce at least a poster for use with festivals and pre-selling. Desktop publishers and graphic designers are everywhere these days, so the creative expense shouldn't run too high. Printing, on the other hand, isn't cheap — especially for a four-color oversized item like a movie or video poster. Check out smaller promo-sized posters and the possibility of designing your materials in black-and-white or with a limited color scheme. Be sure to do a cost comparison with color printing, as the black-and-white design can often cheapen the image of a production, even when done in an artistic manner.

■ M&E tracks – Industry terminology for "music and effects," M&E tracks involve separating every single sound effect and music cue on an audio track separate from the dialogue in a production. This is used in foreign sales, when your dialogue will be dubbed into another language. The M&E track allows for a character to speak with a foreign dialect, but maintains the effects and music present in a scene. For example, if one actor in your film is speaking to another while they walk down a hall, knock on a door, and enter a room filled with music, you need to separate the dialogue between the two actors from all the other sounds present. If some of the effects are married to the dialogue track — say the knock on the door — when the language is rerecorded, the knock on the door will disappear unless it is present on a separate audio track.

Hire a Lawyer

It's definitely worth the $150-$250 to engage the services of a lawyer before signing a contract. If possible, try to work with an entertainment attorney, as he'll be most familiar with the terms and conditions of a distribution contract, and be able to identify any loopholes a potential distributor has written into the deal. In many areas, locating a film- or video-experienced lawyer is not an easy task. You have a couple of options. First, there are many well-known attorneys specializing in independent film and video law who advertise their services in magazines such as *MovieMaker* and *Independent Filmmaker*. You won't be afforded a face-to-face meeting (unless you board a plane or make a road-trip), but that's not really a deterrent. These guys need only see the contract and the project being discussed, which can be easily handled via traditional and electronic mail. Many of these firms also make special presentations to independent film and video associations (such as the Independent Film Project), offering advice and consultation in a seminar format and private meetings following the event at reduced fees. A list of some of the more experienced entertainment attorneys that work with far-reaching clientele is provided in the Appendix (see page 341). If you want to keep your business local, a second option is to ask local-based bands and performers for recommendations to legal counsel. Many of the terms and conditions of a music-representation contract will be similar to a film or video deal, so these attorneys will have some solid footing to investigate any offer coming your way.

Remember, no two contracts are alike. Do not accept a company's standard offer. Any deal you end up with is as much determined by you as it is by the distributor or buyer. Enter contract negotiations prepared. Carefully research all potential buyers. Decide what terms you can and cannot accept.

With all that said... I have to point out that sometimes (maybe *always* in the independent world of self-distribution), any deal is better than no deal. While you don't want to sign with a less-than-reputable distributor and find your project and any potential profits stolen "legally," you also don't want to ruin a deal by refusing to accept less than perfect terms. Is that extra $1 per copy for your film or video really worth sacrificing a distribution contract that will afford you an audience and wide-ranging exposure? I doubt it. While selling *Killer Nerd*, I was told by another independent producer that it was important at that stage of the game (promotion of our first feature film) to become a known entity, to gather awareness and "critical mass" as an independent producer. I believe this observation to be true. Even though our film was an extremely low-budget, shot-on-video horror-comedy, our less-than-perfect distribution contract (not as much money per tape as we wanted, long payment terms, limited advertising) with Hollywood Home Entertainment put the product in video stores across the country, and helped to generate some of that oh-so-desirable "hype." Successful distribution of your film shows the world that you are capable of producing a product that sells, and that makes this whole process a hell of a lot easier the second time around.

Chapter 25 Summary Points

- A so-called "regular" distribution contract should never be accepted — those only benefit the buyer.

- Check a firm's longevity and reputation in the industry before signing a business contract with them.

- Exclusivity means that only one company can represent your film or video in the various markets. Only agree to this term if that company has adequate marketing reach.

- Try to maintain ownership of your film or video by simply licensing the right to sell it instead of selling all rights to it.

- Monetary advances paid at the deal stage are rare in today's independent marketplace. If you are offered one, take it — it may be the only money you'll see from the deal.

- Get a firm commitment from any distributor on the amount of advertising and promotion (vital to sales and exposure) they plan to dedicate to your project.

- Always insist on the right to inspect, audit, or terminate. If a deal starts to go sour, you want the power to legally check out the situation or pull the plug altogether.

GETTING DOWN
TO BUSINESS –
FOLLOW-UP AND FULFILLMENT

Once the excitement of selling a copy of your film or video dissipates, reality starts to set in. The main point of self-distribution — in addition to finding an audience — is to make money. And to do that, you're going to have to run a business. While this chapter is by no means a substitute for a comprehensive primer on the intricacies of operating a small business, the following are some basic practices associated with the self-distribution of an entertainment product:

Filling Orders

Obviously, when selling your project to a broadcast/cable entity, institution, or other "one-copy" outlet, filling the order will not be an issue. Specific technical and contractual guidelines will specify format and delivery timeframe. However, dealing with video rentailers and retailers is another issue. What is the lag time between receiving an order and filling one? And, how will you know how many tapes to have on hand to minimize this delay? Order fulfillment can become a vexing problem for the self-distributor. The key is to balance inventory (or projected orders) with actual orders, resulting in the least amount of out-of-pocket expenses while at the same time limiting payables. Self-distributors are often "one-man bands," and anything you can do to increase or maintain a steady cash flow will contribute to the longevity and success of your efforts. How do you do this? Well, the studios, large independents, wholesalers, and even many of the small independent distributors use a system of **pre-order dates** and **street dates** to assist in their projection of sales. An ad or promotion will be circulated to all potential buyers offering a new film or video product. On the ad will be an order cut-off (pre-order) date and a corresponding street date. All orders must be placed by that pre-order date to guarantee delivery by the street date (which is usually three weeks later). This is a great system for a large-scale promotion that reaches all the potential buyers in a market at once. At the end of the order date, all orders are tabulated; the film or video is duplicated, labeled, packaged, and shrink-wrapped; shipments are made; and the buyers receive the tape in time to rent or sell it by the street date. The use of trade magazine advertising, a national sales effort, order-processing centers, high-quantity duplication and packaging, and possibly thousands of shipments make this an expensive system — one that is out of reach for the self-distributor — at least on this scale.

A more manageable technique involves order and street dates in a territory-by-territory basis. Here's how it works: Our initial promotion for *Pig* was aimed at video rental stores within a 25-mile radius of our locale. Though sell sheets were used, no dates were printed on these materials. Instead, we customized the sell sheet with a sticker bearing specific pre-order and street dates for that territory. In this way, we could closely gauge orders and limit both the order fulfillment period and the amount of money spent on the front-end of the process. Another reason to use this system on a territory-by-territory basis is that as a self-distributor, you will *always* be selling tapes, so you don't ever really want to close the ordering period on a large scale. Even if you already solicited an area, a new buyer might pop up and you can return to that territory with new dates. This technique also avoids placing inefficient, low-quantity duplication orders. If you sell store-to-store, with no implied order cut-off dates, how do you know when to start making dubs, and what time frame do you give the buyer for delivery? Chances are good you don't have the funds (nor do you want to spend available money purely on speculation) to create 200 copies before you start the sales process. It certainly is customer friendly to stock this kind of inventory and offer immediate fulfillment. But nobody expects this kind of service and it's really not worth the investment at this stage.

Duplication/Replication

Making copies of your project on tape or DVD is again mostly a concern for sales to the retail and rental markets. Depending on the project, and the number of expected or realized sales, duplication of VHS tapes or replication of DVD copies can be performed in a variety of ways. Very targeted, niche programming may experience limited, almost "trickle" sales. Duplication in this case can usually be handled by a local production house with a small rack-dubbing system. You'll want to talk with a sales rep to get price sheets, turnaround times, and the availability of ancillary services such as packaging, labeling, and shrink wrap. DVD replication is not always available on a local basis, so check the Internet for providers willing to work with small-quantity jobs. If you find, after initial solicitations, that you'll be moving large quantities of tape or discs, investigate nationally based duplicators and replicators. These companies have the ability to fill big orders quickly, insert the tape or disc into a package, shrink-wrap it, and even drop ship to a buyers' location. Perhaps the biggest benefit of working with a duplication facility is its ability to offer credit. This means you can place an order for 500 tapes and not worry about paying for the product until 30 days after the job is complete. (Note: You should ask any local firms about this option, too.) Terms allow self-distributors to "float" payables long enough to collect receivables. *Killer Nerd* was bulk duplicated and packaged at a firm in Delaware, with some large shipments going directly to Hollywood Home Entertainment in Florida and others shipped to our office, where the tapes were repackaged for smaller deliveries to markets for which we maintained sales rights.

When talking to duplicators and replicators, never accept their initial quote for business. They will always lower the price (unless your orders are a very insignificant number). Ask about other titles they've duplicated and, if possible, rent these from a local store. Look for signal clarity, tape quality, and good sound. Ask for references, specifically independent film and video producers. Call these folks and ask about billing, timely delivery, customer service, quality of product, and price competitiveness. Don't slack on this step and figure a dub is a dub. Your product is your business. Send out a bad tape or a DVD that's incompatible on 50% of the players in the market and you'll never do business with that buyer again.

Packaging/Labeling/Shrink Wrap

Basically, the same concerns apply with these services as with duplicators and replicators. The self-distributor will want to balance quality, price, and timeliness of delivery. Self-distributors soliciting the DVD market have the advantage of working with a local printing company. Unlike the specialty die-cut printing required for a video sleeve, a DVD package consists of a case with a small printed card insert, allowing production virtually anywhere (even on a high-quality desktop printer). The DVD jewel cases are very cheap — less than 25 cents apiece — and can be ordered in a variety of colors and in small-to-large quantities from hundreds of suppliers nationwide.

Selling to the videotape market requires the design and production of a cardboard video sleeve, which is not a cheap proposition. A printer must be selected before you design the package, since each uses a slightly different layout template for final printing. It may seem insignificant, but a 1/16" inch difference is a huge issue when it comes to printing specifications. Like the DVD package, you'll need some good photos, artwork, descriptive copy, credits, and a logo to create the layout for the sleeve.

Independent features live and die by their packaging and must meet both market demands and technological specifications in a budget-conscious manner.

With the photos scanned and the rest of the elements created electronically, a graphic designer can put all of the elements together into a visually pleasing arrangement. Find an artistically (and technologically) inclined friend or an up-and-coming artist to perform the work, which could run anywhere from $100-$500 (or dinner and a six-pack if you go the friend route). Be sure your designer is in close contact with whatever printer you choose; files will need to be delivered digitally so the technical specs are very important. If the printer needs to tweak the design to fit a template — that costs money. Nobody works for free. You'll want a proof, also an additional cost. Most printers won't print less than 100 sleeves at a time, though there are some specialty low-run printers that will deliver smaller quantities. The drawback: cost. At less than 100 units, you may pay upwards of $3.00 a sleeve. If you are selling your tape for less than $15, profits start to get thin with these kinds of costs. To some, it may seem ridiculous to print 250 sleeves. But let's say you originally only printed 50, and then sell 47 copies of your film or video, after which you get a new order for 10 copies. Are you going to tell those buyers they won't get product for a month because you'll be printing new video sleeves? That's not too professional.

Though you won't need labels for a DVD, you will for VHS copies. Your printer should be able to create VHS videotape face labels in addition to the cardboard sleeves. They're cheap, less than a dime apiece, and are necessary for proper identification and copyright of a project. Another option: screen-printing. Instead of using a paper label, the same information can be printed directly on the plastic video cassette body. A layout is designed with the same size specifications of a label, and the tape shell is printed either before or after duplication. Most large duplicators offer this option; you would only want to go this route if your duplicator can handle the job. It would be cost prohibitive to duplicate your program with one facility and screen-print it with another.

THE WIND

Love comes in many forms.

© 1999 Mean Time Productions

While a variety of designs, layouts, and colors can be used when creating VHS face labels, basic information such as program title, running time, and copyright should be included.

T. Michael Conway & Mark Steven Bosko PRESENT

PiG

COLOR • 86 MINUTES • STEREO SURROUND

WARNING: Federal law provides severe civil and criminal penalties for the unauthorized reproduction, distribution and exhibition of copyrighted motion pictures in any medium (Title 17, United States Code, Section 501 and 506). Federal Bureau of Investigation investigates allegations of criminal copyright infringement (Title 17, United States Code, Section 506).

Licensed only for non-commercial private exhibition in homes. Any public performance, other use or copying is strictly prohibited.

2000 T. Michael Conway Mark Steven Bosko Productions, Inc.

VIXEN HIGHWAY

Written, Produced & Directed by John Ervin
prolix@earthlink.net
www.vixenhighway.com

Copyright c 2001 Berlin Productions. All Rights Reserved.
4:3 NTSC VHS   TRT: 75 minutes   12.1.01

Shrink-wrapping is usually done at the duplicating facility as well, unless you are processing very short runs. Again relatively cheap — as little as two cents a package if working with large quantities — shrink-wrapping ensures delivery of a clean video or DVD. A do-it-yourself shrink-wrap kit can also be used. Basically a blow dryer, hot iron, and baggies, this set-up runs about $200 and comes with enough supplies to wrap several hundred tapes or discs. Whatever option you choose, shrink-wrapping also helps the image of a product, presenting it in a professional manner.

Replacements and Returns
Policy issues are a big part of doing business, and two items that buyers will always inquire about are replacements and returns. What they want to know is how do you, as a self-distributor, handle defective or unwanted and unsold merchandise. To be honest, I wasn't truly aware of the impact of these decisions until well into the sales process of *Killer Nerd*. We had shipped 200 copies of the film to a national music and video retail chain. Since I really wanted to land the account, I sold

the tapes at a very low wholesale price, meaning our profit margin on the order was pretty slim. In other words, at that point I was more interested in getting the tape in stores and in front of customers than I was in making a buck. Because of the MTV connection (as mentioned earlier, the lead actor was a part-time MTV VJ), the tape sold fairly well. What I didn't realize, however, was that regardless of the popularity of a product, there are always returns. Okay — maybe *Star Wars Episode One* sells out, but on an independent level, this is not the case. The retailer isn't going to take the hit on that leftover merchandise, so they want to return it for credit. When I got a batch of tapes back in the mail along with a return document, my first reaction was to send them right back again and say, "You bought them, they're yours." Obviously I didn't, and realized I'd have to suffer the loss of money it cost to produce and ship that product. Even worse, the next order sent was almost at no charge because of the credit owed the company.

This is not a problem solely restricted to multiple-unit sales. Another humbling experience with returns occurred during the solicitation of *Pig*. In the middle of our regional video rentailer promotion, I got a call from a very small store asking for his money back on the tapes. Since his business was only about 10 miles away, I decided to stop in personally to discuss the matter. The owner of the store proceeded to tell me that several of his customers thought the movie was so bad that they wanted their money back from the rental charge. His argument was that he only wanted to offer audience-pleasing movies to his customers, otherwise they would stop frequenting the store (and this from a man who had two copies of *Bedazzled* on his shelves). He handed me the two copies of *Pig* (with the video sleeve torn half off one, even) and I handed him a check for $50.

Long story short: Count on a percentage of your sales to come back in the form of returns, and factor this information into any pricing decisions. Also, clearly state your return policy in any sales materials. You can deny returns — just don't count on sales to large retailers; they won't deal with a no-return policy. And, any unhappy present customer may not be so approachable when you solicit your next project. If you are afraid of a large number of returns, you may want to take a second look at your project before approaching the market, as such insecurities may be pointing to the fact that your film or video, in reality, is not up to snuff. Another return policy approach: Offer a free copy of another film or video in addition to the product already shipped. If you only have the one film to offer, you may want to partner with producers of like-product for this practice. Say you've made a deck-building instructional tape, and a buyer would like to return two of the 10 copies purchased. In lieu of the return, you might be able to ship a couple of copies of a landscaping instructional tape from a different producer. You can set-up a reciprocal relationship which allows for your tapes to be exchanged for the other producer's returns as well.

Handling replacements is a no-brainer. You must replace, free-of-charge, any defective tape or DVD sold. Don't charge for shipping or handling. Don't make the buyer return the defective tape. Don't make the process difficult. A simple phone call, e-mail message, or note by mail from the buyer explaining the problem should suffice. In the nearly 15 years of selling and distributing videos, I've only had four defective tape-return situations. If your duplicator/replicator is doing his job, this will not be an issue.

UPC Coding

The Universal Product Code, better known as a UPC, is literally on every product bought and sold in the United States, and elsewhere, today. Information provided by the series of lines and numbers allows global trading business partners (in other words, distributors, producers, and video retailers) to know exactly what is being sold, in what quantities, and to whom. The code allows these businesses to track everything from product inventory levels in a warehouse to how many units a certain film or video has sold at a single location of a national mega-merchandiser. Many buyers, especially large retailers and online sellers, will require a UPC on your product before they even consider stocking it. The proliferation of product and the fact that so many customers use automated order and delivery systems make it simply too difficult to track items by title anymore.

If you decide you need a barcode for your product, you will want to contact the Uniform Code Council (UCC) and ask for an application. They offer a very comprehensive and easy-to-understand Web site — *www.uc-council.org* — that provides explicit details about what you need to do to get started. Also, their customer service system is very responsive and quick to return calls and e-mails on even the most basic questions. The UCC will give you a series of numbers, essentially a pre-fix that identifies your company. Following that portion of the code is another set of numbers that identify your specific product. In this way a company code never changes and any new products offered just get different end numbers.

The big downside to using the UPC is cost. It runs approximately $700 to register for a code, not counting the extra expense involved in getting the bars and numbers on all your packaging. If you've already designed and printed video or DVD packaging, the UPC can be applied with a label. Special software and label sheets are available that work with almost any computer/printer configuration for this specific purpose. It's best to integrate the UPC into the original package to avoid readership and adhesive problems associated with labels, so this decision should be made as early in the promotion process as possible. Though the high price tag might seem a deterrent, you have to weigh this cost against the amount of business you could lose without the code.

It's probably not a bad idea in your survey of potential buyers to inquire as to their use or requirement of UPC in sales and placements. Any project with appeal to a wide audience (feature films, exercise-instruc-tion tapes) will definitely need the UPC for penetration of a national market, and should have the code available before soliciting vendors. There's nothing worse than getting close to a big deal and having it fall apart because of some little detail you overlooked. I knew about the UPC dilemma when soliciting *Killer Nerd*, but held off purchasing the code for budgetary reasons. Midway through the promotion, we found a buyer seriously interested in stocking the tape in all of his stores nationwide. This was a big chain — more than 400 stores — and they were talking about buying 4-6 copies per store. Needless to say, we were ecstatic, until he casually asked about our UPC numbers. Without the code, the deal soured. I tried to talk the buyer into waiting until I applied for and received a code, plus whatever time it would take to print labels and affix them to the huge tape order. But he was already turned off by what appeared to be our unprofessional "rinky-dink" operation. Lesson learned.

Shipping and Freight

This item sounds pretty mundane, but there's more to running a self-distribution operation then throwing a DVD in an envelope and dropping it at the post office. Most buyers you deal with, from the single mom-and-pop video shop to multi-store retailers, will want shipments of product via United Parcel Service, FedEx, or other courier services. The reason? These shippers deliver at nearly the same time every day, rarely lose a package, and offer the ability to track packages. Storeowners can rely on quality delivery service, a major concern with packaged-goods sellers.

With regards to shipping terms, there's really no standard operating procedure when you are self-distributing. When I make a solicitation, I always include shipping in the price. If I advertise a film to video stores for $20, that's what they pay. Obviously this cuts into profits, but small factors such as free shipping and quick order turnaround often help convince a buyer on the fence to order a tape or two. I have charged the buyer for freight only with large retail accounts, where shipping can cost $100 or more per order, and when shipping to foreign accounts — due to the increased shipping and duty costs and paperwork associated with moving videos and DVDs from the United States to Canada or Mexico. You can't simply mail a box of movies to our northern neighbors. Specific regulations governing the content of the film require the completion of forms for immigration passage. This is one time you may want to speak with your neighborhood postmaster to learn of the easiest method of delivery.

Other shipping costs to consider include the price of boxes and cushion filler, padded envelopes, labels, and packing tape. These minor essentials can add up quickly if bought in small quantities. Gauging future sales will allow for cheaper bulk purchases of these and other expendable packing supplies.

Business Entity

Many independent film and videomakers set up an official business entity before beginning production of a project. This allows for the influx and outflow of funds through a company in lieu of personal spending. Better cash control, an accounting record of business, and protection against liable and damage lawsuits are all benefits of running a production professionally instead of as a back-pocket hobby. The same theory applies, and is probably even more important, when self-distributing the finished project. Chances are good that more money is coming into the business at this point (or so it is hoped) than during production, so you need a way to handle incoming funds. Let's say you sell a couple of tapes to a local video store. If you know the proprietor, you might be able to get away with him handing you a couple of twenty-dollar bills or even writing a check in your name. But what happens when you start soliciting out-of-area and out-of-state stores, not to mention large retailers? Do you think the video buyer for Wal-Mart is going to feel comfortable asking his finance department to cut a check made out to a private individual? There's a whole process of purchase orders, payment terms, and credit applications that are set up for business-to-business trade. While it'd be great to have that cash in your pocket, selling tapes or DVDs to a national retailer is not like schlepping videos at a garage sale. Beyond these practical reasons, presenting your product from under the umbrella of a business helps present you as a professional able to compete in the industry. Even if you are working from the basement in your home, a buyer 2,000 miles away has no idea of your physical operation. All he is ever exposed to are your marketing materials, Web site, and any phone/fax lines. There are thousands of successful and apparently well-staffed and officed "companies" doing business that are little more than two guys with a computer, several phone lines, a bunch of boxes, and a stack of tapes and DVDs in a spare bedroom. It's the image that counts in this industry. Setting up your self-distribution efforts as a legal business entity is a must.

The type of business entity you choose pretty much depends on the type of operation you've been running up until this point. One-man bands will obviously want to become "sole proprietors." A **sole proprietorship** is a business that is owned by one person (and sometimes a spouse) and isn't registered with the state as a corporation or a limited liability company (LLC).

Sole proprietorships are very easy to set up and maintain; you may already own one without knowing it. For instance, if you are a freelance videographer or writer who takes jobs on a contract basis and aren't on an employer's regular payroll, you are automatically a sole proprietor. However, you still have to comply with local registration, license, or permit laws to make your business legitimate. And you are personally responsible for paying both income taxes and business debts — the biggest downside of this kind of business structure. A sole proprietor can be held personally liable for any business-related obligation. This means that if your business doesn't pay a supplier, defaults on a debt, or loses a lawsuit, the creditor can legally come after your house or other possessions. So, if you front-end purchase 200 tapes, ship the order, and then get stiffed by the buyer, you personally are responsible for the debt.

A simple self-distribution arrangement for those operations with more than one person is a **partnership**, the simplest and least expensive co-owned business structure to create and maintain. However, there are a few important facts you should know before you begin. First, partners are personally liable for all business debts and obligations, including court judgments. This means that if the business itself can't pay a bill — for example, to a tape supplier, a lender, or even the landlord of your product warehouse, the creditor can legally come after any partner's house, car, or other possessions.

Second, any individual partner can usually bind the whole business to a contract or other business deal. For instance, if your partner signs a contract with a replicator to create a certain number of DVDs that you later find out you don't need and can't sell, you can be held personally responsible for the sum of money owed for the creation of that product under the terms of the contract. Finally, each individual partner can be sued for, and be required to pay, the full amount of any business debt. If this happens, an individual partner's only recourse may be to sue the other partners for their shares of the debt. Because of this combination of personal liability for all partnership debt and the authority of each partner to bind the partnership, it's critical that you trust the people with whom you start your self-distribution business.

Forming a **corporation** limits your personal liability for business debts, but running one can be complicated. Most people have heard that operating a corporation provides "limited liability," that is, it limits your personal liability for business debts. What you may not know is that there's more to creating and running a corporation than filing a few papers. You'll need to keep excellent records to handle the more complicated corporate tax return, and in order to retain your limited liability, you must follow corporate formalities involving decision-making, meetings, and record-keeping. In short, you've got to be organized. And you've got to enlist the services of an attorney to set the whole thing up and keep it running.

Limited liability companies combine the best aspects of partnerships and corporations, offering many benefits to the self-distributor looking to set up shop for the long haul. Those utilizing a limited liability company (LLC) structure will enjoy the corporation's protection from personal liability for business debts and the simpler tax structure of partnerships. And, while setting up an LLC is more difficult than creating a partnership or sole proprietorship, running one is significantly easier than running a corporation.

Self-distributors can now form an LLC with just one person in every state except Massachusetts. While there's no maximum number of owners that an LLC can have, for practical reasons you'll probably want to keep the group small. An LLC that's actively owned and operated by more than five people risks problems with maintaining good communication and reaching consensus among the owners. You don't want to spend your time arguing about the color of your film's logo on the packaging — you want to sell the thing and move on. But if you have a big group of artistic-minded people (such as film and videomakers) acting as members of an LLC, this is the risk you'll take.

Like shareholders of a corporation, all LLC owners are protected from personal liability for business debts and claims. This means that if the business itself can't pay a creditor — such as that tape supplier from the earlier example — the creditor cannot legally come after any LLC member's house, car, or other personal possessions. Because only LLC assets are used to pay off business debts, LLC owners stand to lose only the money that they've invested in the LLC. This feature is often called **limited liability**. Again, you'll need the help of an experienced business lawyer to set-up this form of business entity.

Business Location

As a self-distributor, you won't experience much (if any) walk-in sales, which means concerns for visibility and pedestrian traffic won't drive the location decision. Most first-time self-distributors' first choice for a location is their own home. A fairly empty basement, spare bedroom, or unused garage are all good choices to set-up shop — as long as they are dry, secure, and provide electric and phone line access. Space for storing DVDs, videos, and packing materials, a work area for building orders, the ability to receive shipments/mail, and a traditional desk for making sales calls, writing letters, and keeping track of the business are other considerations.

Running a business from your home is cheap, but the intrusion into your daily routine is sometimes not worth the cost-savings. Unless the room or area is well separated from the rest of the home's living space, it's impossible not to mix business with, well, living. I had a friend who operated his production company/self-distribution business from his home. Though he and his wife didn't have children, the spare bedroom was already occupied as a sewing/sitting room, so he was forced to move the operation to the basement. It didn't take long for boxes of tapes to begin appearing in the living room, the laptop to become per-manently located on the kitchen table, and the home's bathroom to a convenient stop for the UPS delivery driver (who after a couple months of very frequent deliveries and pick-ups, became a personal friend). Needless to say, the businessman's spouse wasn't too pleased with the situation. This is a real concern, because it's inevitable that the actions of doing business will creep into every aspect of normal life. If you live alone and are obsessed with "making it," this is not always a bad thing — just food for thought.

If extra space is not available at your home or apartment, or you do not want to mix business with your personal life, check out low-rent ware-house/office space. Many realtors offer such room in small incre-ments, with five or more businesses operating virtually from the same building. Again, you don't need much beyond technology and security when getting started. Look for space outside a city, near residential or rural areas. It's cheaper, makes for less traffic hassle, and — as opposed to crowded downtown business space — often affords room to grow if you need it.

Banking

Depending on the business entity, you'll need a separate business checking account or at least a **D.B.A.** (doing business as) notation on your account. You never want to mix personal and business funds. Ever. Sure, it costs more to hold two accounts, and there's no such thing as free checking in the business world. But you'll want to project yourself as a professional in every manner. Call several local banks and inquire as to their rates, policies, and requirements for arranging a business checking account. Also ask about credit-card processing — something you may not need immediately, but might before long, depending on your involvement with Internet selling.

Chapter 26 Summary Points

- Self-distributing a film or video means you are going to have to set-up a business.

- Utilizing a system of ordering and shipping time periods allows you to better manage inventory and control costs.

- Duplication, packaging, labeling, shipping, and freight are the nuts and bolts of doing business with retail and rental stores, and demand close attention to the details if you want to offer the best product for the best price.

- Doing business as a self-distributor means swallowing the costs of replacing defective product and return allowances for unsold tapes and DVDs.

- Many large retailers and online sellers require UPCs — those black-and-white barcodes — printed on any product sold through their operation.

- A limited liability corporation (LLC) offers the self-distributor the greatest benefits when it comes to organizing your operation into a legal business entity.

- Keep your business and personal lives separate — physically and financially — to better the chances of both thriving.

USING THE INTERNET – A SPECIAL CASE

Probably the biggest boost to the promotional efforts of independent filmmakers since the invention of video is the Internet. Heralded (and rightfully so) as a cost-effective and efficient sales tool by product pushers of all kinds, the Internet has become especially useful for fledgling film and video artists looking for audiences, attention, and earnings. This technology's role in the success of _The Blair Witch Project_ is by now legend, and legions of other independent producers looking to launch their careers have made use of the medium in a similar manner.

The benefits of using the Internet as a promotional medium are so numerous — low-cost, instant access, ease-of-use, non-invasive, constantly available... it's hard to know where to start. One thing is certain, however, the Internet should not only be considered during the postproduction promotional phase, but also implemented as early in the process of making your film or video as possible. Many projects without a frame shot or even a dollar of financing found, host worldwide Web sites advertising their existence. This kind of proactive publicity can lead to financing, crews, media coverage, even distribution deals, and the best part: It's really pretty cheap.

With the Internet so much a part of our daily lives, this electronic portal is a natural outlet in which to "advertise" your project. Where else could you reach a worldwide audience for pennies a day? In what other medium could you deliver not only images and words, but also actual _moving_ images and sounds from the project itself? Where else but the Internet could you sell your film or tape via credit card or electronic-funds transfer, taking advantage of impulse purchasing? Finally, what other promotional technique allows for instant interactive feedback from fans and customers, information that can help guide your future efforts?

Obviously, the Internet is without equal when it comes to providing a means for the self-distributor to promote, publicize, and sell a project. The first step in taking advantage of the incredible technology is creating a Web site.

BUILDING A SITE

There are two ways to do this — on your own or through the services of a Web developer (either an individual or firm specializing in the service). Of course it's cheaper to take the do-it-yourself approach, and various software programs on the market make it fairly simple. Be aware, however, that many of these "cookie-cutter" site-building tools won't give you the ability to create a cutting-edge Web site that savvy Internet users demand. You will get a presence on the Web though, and that's a priority.

If you have the budget available, you might want to enlist the services of a Web developer. These folks are professional designers with the technical knowledge needed to create a Web site that fulfills your promotional, sales, and distribution needs. Literally thousands of Web developers are in business around the country, and this is one service that can be contracted without a face-to-face meeting. Electronic files, photos, video, sounds, and other information can be transmitted via traditional or electronic mail to the developer, and approval proofs of your site can be returned to you in the same manner.

A hybrid technique mixing do-it-yourself with for-hire services can be found through many of the online Web hosting and domain name registration companies. One such company is VeriSign (*www.verisign.com*), which allows users to "build their own Web sites" using VeriSign's templates and coding. Users choose from a variety of backgrounds, colors, fonts, and artwork, as well as Web site package size to develop a site that best suits their individual needs. For self-distributors, this is a quick and easy way to get information about a film or video on the Net. Advanced packages from the company even allow for online ordering and chat rooms for as little as $150 a year. And yes, that price includes the hosting and creation of the site. Try to match those rates with any other advertising medium and you'll see why the Internet is such a must-use tool.

The benefit of building your own site is that you are the person most knowledgeable about your film or video, have the most passion for the success of the project, and will work harder than anyone else in promoting it to the world. The downside is learning new technology and acquiring skills (such as layout and design) that may at first seem foreign (not to mention technologically frustrating). There's also the time issue. If you are holding down a full-time job in addition to fulfilling all the duties associated with self-distribution, do you have a couple of extra hours in the day to devote to learning Web design?

Whatever route you choose, the main reason to host a Web site for your film or video is to assist in the promotion, distribution, and sales of the product. In other words, you want to be sure your site is doing two things: getting traffic and getting that traffic (the consumers who have come to the site) to take some action; otherwise there's not much reason to be on the Web.

Depending on what stage you are in the promotional process and what audience you are appealing to, this Web-based action can take many forms:

- Further inquiry
- Distribution offer
- Request for full-length screener
- Invitation to interview
- Simple "fan" correspondence
- Continued dissemination of your information
- Sales

Like any advertising, publicity stunt, promotional campaign, or solicitation, your Web site should first attract the desired audience and then call that "viewer" to action. Even if it's just to watch a trailer, elicit feedback on a guest book page, or instigate further interest in the project or you personally, your site needs to intrigue and involve anyone stopping by for a look.

The Web site for *Pig* was developed at the height of the *Blair Witch* craze, and, as expected, took on many of the properties that made that blockbuster's Web presence such a success. Opting to have the site designed by a professional (we found a designer just starting out who agreed to create the whole thing for $400 plus a copy of the film), the *Pig* site reflected the reality-basis of the movie, asking the viewer to decide if our film about a small-town cop was actually real or just a fictional creation presented through a series of documents and other bits of information supposedly "found" during the investigation into the title police officer character.

One of our goals was to provide Web surfers with a unique promotion for what was ultimately just another independent film trying to make it big. We wanted more than just our bios, headshots, and an order form for the movie. The design of the site attempted to create its own environment and give the visitor a feel for what it was like to live in *Pig's* world. Basically, we were aiming for something neat and cool and different.

The goal of the *Pig* site, however, while aiming to please potential viewers and attempt to start an underground, low-budget buzz on the flick, was primarily to present the film as a viable product to interested buyers. We didn't want to come off as some hobbyist filmmakers who scrapped together a hundred bucks and a video camera for a backyard stab at the craft. Our Web presence not only defined the genre and feeling elicited by the production, but also served to elevate the quality of the product itself by presenting new and exhaustive information (much of which isn't found in the film) that perpetuated the reality aspect of the movie in a professional and first-class manner.

For example, *Pig* was designed to give the viewer quick glances into the many episodes that make up a small-town policeman's daily life. In doing so, many characters are never identified by name; they are merely arrestees in a long day and night of the policing job. On the *Pig* site though, there is a detailed description of each of the characters that includes names, occupations, age, reasons for arrest, and sometimes a quote or two on how they felt about the title character and his manner of work. In one scene, an obviously mentally challenged man is arrested while reading his manifesto about life. He is forcibly removed from a barbershop, scattering his papers as he goes. Visitors to the site could read the words and thoughts from these pages, as each was reproduced in its entirety on the Web.

Also on the *Pig* site was a media area which allowed for the downloading of news releases and photos, a contact section, visitor forum, trailer screening "room," and even a dictionary of the slang policeman's terminology frequently used in the movie. We wanted to make the site as useful as possible to both the media and any interested distributors, and launched it just prior to completion of the finished film.

Of course, every film or video project will have different objectives and topics that can be exploited via the Web, so a good way to get started on building a site is to brainstorm and write down all the ideas you have that would lead to the successful promotion of your project. Think about your audiences and potential buyers in deciding what content is most appropriate and would serve you the best in achieving any sales or distribution goals.

We all like to see ourselves "on screen," but is it vital to have a section of your site dedicated to you, the creator of the project? Possibly, if your expertise lends credibility to the project. But if this is your first independent feature or instructional video, and you have no pertinent background in the topic presented, a long-winded bio really isn't necessary. Remember, people buy products. Now, just as with the written public relations and publicity tactics, if you have a unique personal story to tell that ties into the production or release of the film or video, by all means, tell it! If you financed your feature by convincing the entire cast to give plasma donations twice a week — that's interesting to many folks and says a lot about your chutzpah in pushing a project through completion. Distributors like those kinds of hooks; it may help to serve as that extra punch in the acquisitions decision.

Be sure to include both a short (five sentences or less) and long synopsis of the film or video on the site. At first, we resisted this inclusion, arguing that a synopsis wrecked the illusion of *Pig's* reality. After a couple of weeks we investigated the guest book messages and posts to the site, and found that many of these messages, though complimentary, asked what the movie was all about. Visitors wanted to know the storyline. And these responses didn't just come from the occasional surfer — some of the synopsis requests came from distributors (who had been solicited via a postcard) visiting the site. You can be sure we got that synopsis up on the site immediately!

Finally, even the simplest Web site needs a contact page. There is nothing more frustrating than to hunt around a site trying to figure out how to contact the people in charge of what you are viewing. Make your contact area easy to find. Don't simply list e-mail addresses, but also street or P.O. Box addresses and phone numbers, as many distributors and buyers like the old-fashioned method of picking up a phone and calling someone.

When creating a site, don't feel that everything you do has to be original. Don't copy, but look to other film sites, as well as distributor sites, for ideas. What about these entertainment-industry sites do you like? What is appealing? What makes them easy to use? Why would you buy something from these sites? Why do you like the way they present information. Adapt the good ideas you find to the presentation and promotion of your project.

ELECTRONIC PROMOTION – PART 1

Using the Internet is easily one of the most efficient methods available to promote your film or video to target audiences before the actual sales period. Creation of a customized Web site gives you a medium to present your project to the world in order to create interest and anticipation for the film or video when it is available for sale. One problem: The world is a pretty big marketplace. How will you approach and direct this sizable group of potential buyers to your site in a manner that both makes sense and works?

Well, a first approach would mimic the traditional direct-mail method of mailing a solicitation to electronic consumer "addresses," bought through services according to specific demographic or interest profiles. Let's say you've created a site for the promotion of your exercise instructional tape. To drive targeted traffic to the site, you can purchase a list of e-mail addresses of people known to be interested and participate in athletic pursuits. Specialty-list sellers market these names and addresses as well as other media that cater to the same audience. Most special-interest magazines also now collect e-mail addresses when taking subscriptions, and offer this information for sale as well.

But sending a batch e-mail solicitation to a group (known as **spam**) is the equivalent of junk mail, and is often dealt with in the same manner — the material (in this case an e-mail) is immediately placed in the trashcan. Unless the recipient is genuinely interested in the appeal, it can be a colossal waste of time (and funds). So how can you utilize an address list without offending the receivers? That's a tough one. People have become so suspicious of any e-mail that isn't from someone they recognize, the messages are often discarded without opening. I know it's tempting to immediately use your Web site as a "store," but unless you have something truly unique that can't be found anywhere else, it's nearly impossible to sell to the general consumer in this manner.

A better approach is to start your promotion one step up the buying ladder, gathering e-mail addresses for video buyers and distributors. This audience is interested in buying product, and if you can facilitate the process for them in any way, it usually helps the sales effort. Offering product information via the Internet is much less cumbersome and intrusive than standard mail, and it can't get lost or thrown away like a paper package. You can send a quick e-mail to the buyers/distributors, alerting them to your project's status (available for sale, coming soon, half-price, etc.) and include a link to your site. We used a combination of the two media (standard and electronic mail) to alert distributors and other potential customers to *Pig's* impending availability. A postcard bearing the film's key art carried a handwritten note asking the recipient to visit the site for more information. This technique worked well because it allowed the buyer to hold and look at the main marketing material (essentially the video sleeve) while at the same time giving them instant and at-their-discretion access to the comprehensive information on our site.

This same approach works well with any media contacts that might feel overwhelmed with an inch-thick media kit sitting on their desk. This group is especially responsive if you make their job easier, and now press releases, cast and crew information, synopses, art, and other details pertaining to your film or video can just be a click away.

At one point early in the promotion process, we became so dependent on the *Pig* site as a source of information, we simply sent a card printed with the Web site address to media interested in covering the project. Reporters and writers could go to the site and select exactly what they wanted to pursue from the numerous sections of data on the film. There was even a form available to request a full-length screener (the trailer was already on the site). It would have been cost-prohibitive for us to reproduce all of that material, much less mail it to every media outlet we solicited for coverage. This kind of correspondence also allowed for the development of an e-mail connection with the media. Writers and reporters are not always willing to give out their e-mail addresses — especially to someone looking for publicity. By visiting the site on their own, however, most media that covered the film also sent us a note or some follow-up questions via the e-mail link on the site. They didn't feel pressured to provide contact information until it was on their terms. And we didn't abuse this contact by e-mailing the writer or reviewer every two days. Only when something significant happened with the movie (such as when one reviewer compared it to *Taxi Driver*) did we "mail" an announcement to our electronically available press friends.

ONLINE SCREENING

As a self-distributor promoting a film or video, without question the most accessed section of your promotional site will be the trailer/clip area. All the photos, logos, artwork, reviews, and stories are interesting, helpful, and necessary to maximize your promotional effort. But it's the product that buyers, distributors, media, and the general public will want to see before making any decisions. Offering a clip of your film or video for review fulfills this need.

Though a slew of technical considerations must be worked out before placing a trailer on your site (questions as to file size, format, encoding, etc.), all of which can be handled by the hosting company, the one necessity is making your video available in a variety of bandwidth formats. Unfortunately, not everyone surfing the Internet is hooked up via a cable modem. This means that some users have old-school 56K models, others are operating at 128k, and some will access your site via a T-1 line (usually through their workplace). In addition to the trailer, many sites are now offering select scenes, behind-the-scenes footage, outtakes, and even cast/crew interviews as viewing options. What to include really depends on what you are trying to do with the site. We didn't feel the need to show scenes from *Pig* online, as we thought the trailer said enough. And since we're not anybody all that interesting, interviews weren't an option. If I was doing a film site to appeal to new independent filmmakers now, I might include some interviews and behind-the-scenes footage. But that was not our target audience and we didn't want to give buyers anything more than a trailer, which would hopefully wet their appetites for the full-length screener.

Web-based film festivals are another way to electronically promote your project. Though you are not ultimately in control, the forum provides access to a well-targeted viewing audience (that is defined by the subject matter and parameters of the festival) as well as potential distributors. Like traditional fests, many online festivals make it a point to include distributors, agents, video buyers, and other industry professionals as part of their "panels," increasing awareness of the winning entries within an important group of decision-makers. The movies and videos are available either for free or a fee, for a limited amount of time through the festival site. An extreme example was the Sync, one of the Web's first film festivals, which asked viewers to vote for their favorite shorts in seven categories on a daily basis. New films were added each month, and there were even new winners every minute. Audience participation determined which films remained on the site, while less popular films were periodically replaced with new entries.

Many online fests have since evolved into a hybrid between a festival experience and a revenue-producing streaming site, offering independent film and videomakers that longed-for distribution opportunity previously unavailable through other media.

DIGITAL DISTRIBUTION
Can you self-distribute a movie online? If it means using the technology to get a full-length version of your film or video in front of a paying audience, then the answer is yes. The Internet also allows you to advertise your project to potential buyers, much like a traditional distributor offers catalogs and sell sheets when soliciting video retailers and rentailers with new product. And the Web can be used to place and process orders, book screenings, track shipments, and accept payments from

buying customers. Finally, the Internet has no boundaries, meaning foreign, institutional, PPV, cable, and broadcast buyers — not to mention private individuals — are all within your reach, in addition to the video stores and electronic retailers most often solicited by a self-distributor. So I guess the answer to the original question is an unqualified "yes!"

Obviously the easiest manner to "distribute" your movie or video through the Web is with a sales section on your site. Depending on your project and your goals, this can be set up in one of two ways: Give viewers the opportunity to purchase a streaming version of your project (via a download), or sell actual hard copies of the video or DVD (in a mail-order scenario) using some sort of e-commerce arrangement (like PayPal). The streaming set-up is technologically complicated, and usually requires you to align with another online marketer offering the service to film and videomakers.

STREAMING

One of the most enduring of these businesses, ALWAYS independent (*www.alwaysi.hollywood.com*), was started to serve the creators of independent entertainment by developing new revenue and distribution opportunities and to provide an exhibition platform for independent producers to showcase their material to talent representatives, acquisitions executives, and other industry professionals. Self-distributors earn money through the site when viewers choose to download their movie for viewing. ALWAYSi's revenue sharing plan was designed to begin paying filmmakers as quickly as possible. There are no expense deductions to be recouped before sharing begins and the company shares 100% of all subscription revenue with the filmmakers showing films on the site.

Here's how it works: Each month the company tallies up the total revenue generated from subscriptions. They track the number of times every film is watched on the site (although each subscriber can watch each film as many times as she wants, she will only get counted once per film per month) and then calculate the percentage of total streams for which each filmmaker is responsible (your film's number of views divided by total site film views). This percentage is applied to the total subscription revenue to allocate the correct pro-rata sharable revenue amount to each filmmaker, which is then applied to the revenue-sharing rate chosen by the independent producer. ALWAYSi offers a non-exclusive agreement, which gives independent producers 10% of the take, and an exclusive agreement that binds the producers to only using the company for Internet distribution, but rewards them with a 40% share of the revenue.

Will you get rich? Consider the following example that ALWAYSi offers on their site. Say $20,000 in new subscriptions was made in a month. During that time, your project was viewed 5,000 times out of a total of 250,000 film views (this includes shorts and features). If you were on the exclusive plan, you would have made $160 that month. Calculate your revenue share as follows:

> $20,000 total revenue x (5,000 your views / 250,000 total views) x 40% = $160

As you can see, the larger the subscriber base and the more people who watch your film, the more you stand to earn. While you are not required to make ALWAYSi your exclusive online distributor, doing so will quadruple your revenue sharing percentage over the non-exclusive plan (which, in this example, would have earned you $40). Neither figure represents a lot of cash, but it is a way to create a continuing revenue stream as well as get your movie or video in front of a paying audience, and that is the whole point of self-distribution — correct?

Is there a downside to streaming your film on the Internet in this manner? Well, some distributors may tell you that they won't even consider a movie that's been showing on the Net, but if you are at the stage of self-distributing your movie or video, chances are you've already unsuccessfully solicited a good number of those distributors. Film and videomakers who haven't begun the solicitation process should probably hold back and gauge their decision based on the distributor survey. Ask buyers how they feel about the technology and streaming issue. It's certainly understandable — if they're planning on advertising and selling your movie sometime in the near future — that they don't want it available to the public for next to nothing.

There are a number of companies providing this service, some more in tune with independent product than others. Level of service, deals offered, technological requirements, reputation of the provider, and number of subscribers and films/videos offered varies widely as this service continues to grow.

VIRTUAL SELLING

Like a never-ending display ad in a magazine, a section of your Web site can be devoted strictly to the sale of product. On that "page," viewers can order a copy of the project in either VHS or DVD (or as a download if your host allows for this technology), select shipping options, and pay with a credit card, debit card, check, or automatic-funds transfer. All of these e-commerce functions can now be "bought" and installed as services on your own personal site. Web hosting companies offer these shopping services in a wide range of sophistication and prices, but if you are only selling a single product, usually the most basic service will do. Order confirmations will be e-mailed from a third party, leaving you responsible for shipping the product. If you anticipate a large demand or are unable to fill orders on a timely basis, some companies offer fulfillment services. You merely ship packaged product to them and they handle the rest. Operating in this manner will cut into your profits but also free up a lot of time (that may be better spent promoting and publicizing the project). Time and money are always conflicting factors in the self-distribution business.

Products sold through a Web site are subject to any applicable state sales taxes. Be sure to check with local government offices about online retailing regulations and the necessity (if any) of obtaining a sales permit for such activity. If you have established a business entity as described in the last chapter, then these details will have already been worked out. Sole proprietorships need to take special caution in their operation of online commerce as auditors have been paying close attention to home-based businesses profiting from the Web. Be sure to do things right from the very first sale so you don't have to try and backtrack your business dealings for the Internal Revenue Service.

LISTINGS

Another form of online distribution involves listing your product (along with a clip or link to a splinter or private site) on one of the sites created strictly for the business-to-business activities involved with entertainment product distribution. Companies like ReelPlay (*www.reelplay.com*) are not only a great place to do distributor and buyer research, but also provide a forum in which you can present your film or video to the professional entertainment industry. Since its inception in 1999, the company has aligned itself with more 2,200 companies representing 13,260 titles. Producers can build a free film/video Web site that provides acquisitions and development executives, investors, festival programmers, and industry professionals immediate access to all the facts about the project. Paid sites are also an option, and include extras such as art and trailer clips. ReelPlay even offers marketing services to the independent self-distributor. E-mails and/or faxes from Reelplay Blast! are targeted direct mail pieces sent to the company's proprietary database of 20,000 distribution professionals at over 7,500 companies (includes phone, fax, territory, media, genres, and markets attended). For less than 50 cents an e-mail (or $1.50 a fax), this is an extremely affordable manner in which to promote your project. And, by working through an industry-affiliated company instead of sending something out on your own, chances are greatly improved that your promotion will not go unnoticed.

ELECTRONIC PROMOTION – PART 2

One of the most important uses of a film or video Web site is to continue to promote the project to those involved in creating it. Actors, crew, writers, even caterers, have a way of disappearing after a shoot. During a production, the camaraderie between everyone on a set is incredible. You become family, promising to always stay in touch, making plans for future non-work-related get-togethers, beginning relationships that you are sure will last forever. Once the camera clicks off for the last time, however, things usually change. Everyone has another job or an old life to return to, and those close ties quietly slip away.

An active Web site can alter that dissonance, at least electronically. Before people split, be sure to circulate the Web address of your site to everyone and anyone even remotely involved with the production. When it came time for *Pig* to wind down, I made sure even the people offering locations were aware of our Web presence. My goal was to keep this large group of people from ever completely separating themselves from the movie. If they couldn't be reached via traditional mail, and didn't know us well enough to call or be contacted in a personal manner, then at least they could check our Web site from time to time to keep up with what was happening on the film.

Anyone connected to your project is the cheapest and most positive promotional tool you have available. Good word-of-mouth is worth more than any advertising; people love to talk about things they've done, so it is a no-brainer to use these folks as living advertisements for your film or video — and the more they know the better they can promote the project. Keeping them involved, updated, and interested in the project will also serve you well come time for sales. On our site, we listed everyone — and I mean everyone — who had anything to do with the film in any way. We also listed businesses that contributed time and/or facilities and placed links to their commercial sites from the *Pig* site. There was a section on how to contact talent, crew, and other behind-the-scenes people (such as the graphic designer responsible for the key art and the artist who made the wardrobe) and a feedback form for these same people to submit address and contact information changes (actors seem to move a lot). Whenever something big happened with the film (final cut, offer from a distributor, sales into new stores and regions) we posted the news on the site, and alerted hundreds (who offered e-mail addresses) to check out the site for project updates.

If you plan on driving a significant number of the same people to your site frequently, then it's vital to change the information, even the look of the Web page, as often as possible. Think about the sites you surf most often. I'm sure none of them is filled with so-called "stagnant" content — stuff that never changes. The reason a site becomes attractive is because it always has something new to offer. That has to be your goal in maintaining an effective site. Always offer something new. Even if it is as simple as changing the home page greeting, adding a quick news story, or uploading a behind-the-scenes photo every couple days — you have to offer the viewer a reason to return. A good technique to ensure repeat visits: contests. These don't have to be complicated — ask visitors to enter their names for a free movie or video give-away, pose a question about your film or the subject of your video, with the winner receiving a prop or something of value from a business that worked with you on the project — any way to get people to stop and spend some time, and then return, is what you need to do.

Daily updating to a site is also important for those folks you are soliciting. If your Web pages can interest a potential buyer or distributor enough to elicit a return visit, your chances of making a deal are improved. Buyers need to be convinced that there is an audience for the product. By hosting an active site that entertains and informs an audience, it becomes apparent that your film or video appeals to more than just a small audience of wannabe filmmakers.

It's very obvious when a site is designed for and used by a limited number of people. Usually littered with in-jokes and references to events that were only privy to this "inner ring," these kind of sites aren't really on the Web to promote a project as much as they are to feed someone's ego. It does you no good to host a site that becomes an expensive chat room for you and your circle of friends. And don't make smarmy jokes and comments cutting on the entertainment industry, other films, videos, or their creators or your aspiring dreams as a film/video artist. Leave that stuff to review and general-interest entertainment Web sites. If information is not promoting your project in any way, or serving to create some sort of community of supporters, leave it out.

BRINGING 'EM BACK
For those with the time, the ultimate method for finding and keeping return visitors is to create some sort of membership alliance in which a Web surfer becomes associated with your project in some official fashion. You want to build a growing base of fans, like the "KISS army," who wish to publicly align themselves with the film or video. Offer joining members access to restricted password-only areas of the site where they can see and learn more about the project. Special prices on products, freebie giveaways, invitations to screenings, and live chats are all options you can offer to anyone wishing to become a member of your film or video's Web site. These members can create "profiles" of themselves (or alter egos) on the site, listing interests, activities, etc. Basically, a member area is like an expanded guest register, only instead of just leaving a name and a note, you are able to gather

more information on visitors and draw them into the site for more frequent and longer visits. Communicate the benefits of membership and
make profile creation as easy and as fun as possible to keep the member information up-to-date and evolving.

Another way to drum up interest in both your film and site in a non-
obtrusive manner is through the use of an electronic postcard. The *Pig*
site offered a selection of three movie frames that served as the background for the postcard. The sender could enter a message and a
"send-to" e-mail address, using our site as a communication portal.
The postcard automatically took recipients to our site if they so wished.

A quick way to drive traffic to your site: Win an award. In addition to
sending new viewers your way, a Web award can also add credibility to
the site. Most award-boasting sites also boast huge increases in traffic for several days following the announcement. This happens
because the award-presenting entity publicizes which sites have won an
award, and many surfers flock to the honored Web pages to see what's
so special. All types of sites are considered for awards, but the most
common characteristics of winners are unusual design, unique presentation of material, and subject matter appealing to broad audiences.
That sums up the audience quite nicely for many independent feature
films and special-interest videos, so sites of this nature are a shoe-in.
There are many places to submit your site for awards, and if you win
one, you'll be given an icon to place on your home page announcing
the honor. The following represent several good Web-award search
starting points:

- *www.webaward.org*
- *www.websiteawards.xe.net*
- *www.botw.org*
- *www.webbyawards.com*

Here are some other quick ways to promote and increase traffic to your
film or video Web site:

- "Talk" about both your site and its content in subject-specific
  and film/videomaking-oriented Internet discussion newsgroups
  and mailing lists.

- Join discussion groups on the online services (CompuServe, America Online, MSN, etc.) to create electronic buzz about your site, the project, or one of the project's marketing hooks. Have many people associated with the project do the same.
- Solicit and write articles for electronic newsletters (of which there are thousands being distributed over the Internet), adding credibility and authority to both your status as a film/videomaker and the content of the project.
- Do not forget to "advertise" your Web site address on everything you print, including business cards, stationery, envelopes, invoices, advertisements, and any marketing materials and packaging.

ADVERTISING ON THE NET

Using the Internet to advertise your film or video has one huge advantage over traditional means: Much of it can be done relatively free. Okay, I'm sure the word "free" got your attention, and you're probably thinking there's a catch. Well, if you count time as a cost (which it is) then the technique is not totally free. But you can get a lot of attention for your site without any true monetary outlay. And the quickest way to start advertising your site is through the use of search engines.

Google, Yahoo, MSN, Dogpile… you know the names. You visit and use these sites daily to lead you to the information you need; the same function applies to a buyer looking for new film and video projects to explore. Search engines allow a user to enter keywords (i.e., independent film, instructional video), click "enter," and get in return a list of Web sites that offer corresponding subject matter. Being near the top, or even on the first page of this list, is valuable to a self-distributor since many surfers never venture farther than one-click searching. How does a site get priority listing? While there's no foolproof (or free) method to ensure placement on the list, many search engines rely on **meta tags**, specifically the "keywords" and "description" tags, in delivering the users' desired information. These "tags" contain words entered by the Web builder that describe the contents of the site. When a search is conducted, the billions of words in the tags of all existing Web sites are scanned, and those most closely matching the keywords entered by a user are returned in the list.

GETTING IT FOR FREE

So what does all of this have to do with free advertising? Well, if you tag correctly (which is free), chances are good that your Web address will appear in searches for films and videos of your subject category, leading users to your site. For example, by entering the words "independent horror movie" in MSN's search area, various Web site listings appear, one of which is for the independently produced and self-distributed *Satan's Menagerie*. To get your site at the fingertips of a potential worldwide audience that quickly and easily is pretty potent stuff.

When entering tags, list those words, and synonyms for the words, that best describe the contents of your site. Note that meta tags are not the only place for important keywords. Relevant keywords should also be placed in your Web page's title and in the headers of every page of the site. Another trick is to give your page a file name that makes use of one of your prime keywords, as well as including keywords in all of the "alt" image tags. If not building the site yourself, be sure your designer is aware of the keywords needed for each of these techniques. Know that each search engine uses a slightly different procedure in finding site matches, and it's advisable to check out theses services' help pages to see what procedures will get you optimum listing placement. Since you can only do your tags one way, design them to work best for whatever search engine has the potential for driving the most traffic your way.

More free Web advertising can be found by simply listing your film on the many databases created specifically for that purpose. Web sites like The Film Underground (*www.filmunderground.com*) and the Indy Film Network (*www.indyfilm.buried.com*) are just two of the dozens of sites that allow for the free listing of independent films and videos and their corresponding Web sites. Most feature simple forms that ask for basic information about the cast, crew, and the project itself. Finding and sifting through these sites to find those most appropriate will take some time, but that's the only expenditure. Industry and fellow film and videomaker contacts, sales, and just general "traffic" are all outcomes of using these free listing services.

Another way to increase visits to your site without dipping into the budget: Align yourself with a **Web ring**. If someone is searching the Web for a specific topic — like kung fu movies — they'll want to visit as many related sites as possible on the topic. A kung fu movie Web ring is a grouping of various kung fu movie sites, organized so visitors can quickly click to another site within the group that deals specifically with the same subject matter. Again, dozens of independent filmmaking-oriented Web rings exist; all it takes is the time and effort to locate and join those that make sense for your project. It's important to note that there are Web rings for every subject imaginable, and you shouldn't limit yourself to film- or video-only sites. If you did make a kung fu feature, you'll want to investigate any kung fu Web rings out there, as proponents of the sport are also likely fans and potential customers.

Linking to a site that features subject matter similar to either the pursuit of film/videomaking or the content matter of your project will also produce traffic free-of-charge. Similar to the Web ring principle, but not as organized, most Web sites feature a "links" section that allows visitors the opportunity to quickly link to other sites of similar appeal.

Let's say you've produced a wildlife documentary. Any environmentally oriented site provides a good opportunity for hosting your link. Not all Web sites will want to feature your site as a link, but what does it hurt to ask? Those that do may ask for a copy of the tape or DVD in exchange for the "free" advertising, but usually it'll cost nothing more than a request.

Using links often leads to the popular barter system used extensively with advertising on the Web. Though technically still "free," since no money is being exchanged, what happens in this situation is that someone will link to your site if you do the same for them. Looking at the environmental video example above, what may take place is that when you ask for a link to a site that promotes Earth Day, they may want a link on your site. If it makes sense, do it, as being affiliated with other sites — especially large, well-known, or trafficked ones — adds some authority and legitimacy to your production.

After trading links, the next step is to trade ads. **Banner ads**, those commercials rolling at the top and sides of a Web page, are not always paid for. In fact, many (especially those appearing on smaller sites) are traded like links between Web sites that look to attract common viewers. A banner ad is designed to do two things; first, if designed correctly, effectively advertise a project. Most savvy Web users just look at banner ads, clicking only if very interested in the product. This action triggers the second purpose of a banner ad: to increase a Web site's traffic.

Banner ads have standard size specifications and can be created very quickly by a graphic designer or whoever makes your site. Keep the ads simple with regard to the amount of information presented — viewers will only take a quick look before deciding to disregard or take action.

BUYING CLICKS

Obviously, banner ads can be bought; they constitute the main form of paid advertising on the Web. Any commercial site on the Internet that is designed to appeal to a large group of viewers will offer a media kit similar to those of traditional media. In the kit will be demographic data, general site description information, and pricing for the various advertising opportunities. The same factors weighed in print and broadcast advertising decisions must be considered here, including:

- Does your audience "visit" the site under consideration?
- How often does your audience visit the site?
- How much does reaching this audience cost?
- What kind of return on investment will you get for the ad?

A neat advantage Web advertising has over other media is that the advertiser can be billed only for those people actually taking action on the ad. Known as "click-through" or "cost-per-click" pricing (CPC), these ads are paid for by the advertiser only when a Web user clicks on the ad, regardless of how many times it's shown. Google, a popular search engine, offers CPC pricing for ads located on appropriate Web search pages, allowing for a promotional campaign to be started for as little as $5. Their program lets advertisers set their own budget limits before the ad begins its run; once that amount is reached (meaning a specific number of visitors have clicked your ad), the promotion can either be stopped or more funds can be added to continue. All search engine sites and most every commercial site offers some form of click-through pricing. The method is an effective way to monitor traffic to your Web site, as well as measure the effectiveness of the ad and its location on the Web.

Film and videomakers will want to explore not only search engine, film/video, and subject-related sites, but also those Web pages that offer products, services, or information attractive to your customer base. For example, a self-distributor offering a do-it-yourself interior home design video might want to check out advertising on sites that present mortgage refinancing and home loan information. Visitors to those sites often use such funds for remodeling, which is essentially the video's target audience. The same goes for independent feature films. Got a horror feature? Why not promote it on a site that sells Halloween make-up and supplies? Is your movie about fast cars and fast women? Look to automotive specialty supplier Web sites as an outlet for your banner ad.

Don't simply consider the core interests of your audience when deciding on Web ad placement. Unlike traditional print and broadcast media, the Web allows for instantaneous movement from site to site, from subject to subject. A horror-movie buff probably isn't going to pick up anything but *Fangoria* at the newsstand, but he will definitely visit hundreds of other Web sites during a common day of surfing. Keep that mentality in mind as you consider the thousands of options available in the electronic media environment.

Beware of the temptation, however, of promoting (and producing) your site "over" your film. Unlike a movie or video, which may take months to promote, publicize, and sell, the Web offers immediate feedback to your efforts. You launch a site and within minutes people from around the world have access to your information. This kind of instant response can overwhelm what seems to be the slow and unproductive activity of actually selling a tape or DVD. The frustrated film or videomaker then begins to focus his attention on Web maintenance and promotion, excited at the new communication and feedback the medium provides. This is readily evident in many independent film sites. Their level of sophistication obviously outweighs the production values of the project itself. Since Web design is a creative process much like film and videomaking, the emphasis of the work shifts from the film or video to the Web. Requiring the labors of only one person makes this unconscious change of priorities all the easier. Add in the facts that you can work from your bed, don't need to personally interact with anyone else, rarely face rejection, and can communicate with people sharing similar interests and hardships, and you can easily see why this activity can overshadow the hard work involved in correctly promoting a film to distributors, buyers, and audiences. Remember, your priority is to get your movie or video seen by a paying audience. A good Web site is important, but not if it jeopardizes your core activities.

In addition to banner ads, your Web site (or product) can be promoted through the hundreds of electronic newsletters delivered directly to recipient's e-mail accounts. *Film Threat Weekly* is a good example. The e-mail newsletter of the main Film Threat site, *Weekly* is delivered to 90,000+ film fans as well as important members of the film community who read the newsletter. The media kit states that their list of subscribers "contains e-mail addresses from places like: *@newline.com, @warner.bros.com, @ifc.com, @sundancechannel.com, @miramax.com, @paramount.com, @fox.com, @dreamworks.com, @hbo.com, @mtv.com, @disney_studios.com, @cbs.com, @nbc.com,...."* For $125, you can place a 10-word ad (extra words are $12.50 each) that will be included within the body copy of the newsletter, making it harder for the reader to ignore it than a banner ad. Again, look to subjects of both primary and ancillary interest to your target audience.

If an increase in traffic is one of your desires, traditional media provide a good forum for advertising your Web site. When the *Pig* site was launched, a classified ad intended to generate traffic appeared in the Independent Film Project's monthly magazine/newsletter. It featured an attention-getting statement ("Do you hate cops?") and the Web address. Our intention was to create interest among fellow filmmakers and industry professionals by appealing to their curiosity. Once at the Web page, they'd realize it was a promotional site for an independent feature and hopefully explore the many sections. Placements like this are very inexpensive and unique as it's uncommon to see Web sites promoted through low-cost classified ads. Depending on your budget, any and all media can be exploited to advertise your site (billboards have become a popular vehicle for Web advertising), with each offering distinctive benefits. Don't discount any technique available. I once saw a Web site address created from sand on a busy beach during Spring Break. Sure, most people passing didn't have their laptops handy, but the free promotion landed a photo in the following day's newspaper.

CONNECTING WITH THE CREATIVE COMMUNITY

One of the greatest benefits of promoting your project on the Internet is the instant access to "community" which the medium provides. Associations, memberships, and groups supporting every interest imaginable host Web pages, and most have chat rooms, forums, and profile listings specifically for their members. IndieClub (*www.indieclub.com*), one of the first independent filmmaking "clubs" on the Internet, is a portal for anyone with "a passion for making movies using film or video and looking for others who have similar interests." The site is designed for networking with others involved in acting, directing, producing, writing, lighting, or any other skill or desire that is associated with independent film or videomaking. Members fill out profile questionnaires that list their specialties in the industry and any relevant experience, which are then posted so members can contact/correspond with each other. Local in-person meeting groups have been arranged through the site, as well as special discounts on film/videomaking products and services.

IndieClub is a free organization, as are many, but literally thousands of "for-fee" virtual memberships do exist on the Web. Explore joining those that make the most sense for your specific project's subject matter and will benefit you in a promotional sense. Be careful not to get sucked into the whole Internet communication experience, wasting valuable time that could be spent promoting your film. It is interesting, and sometimes beneficial, to "chat" with others about the successes and failures you are encountering promoting your project — I've been led to a couple of sales from such exchanges. Just don't let the electronic chitchat overtake your primary activity of searching for an audience.

FINDING BUYERS

Back in the discussion on researching distributors, the Internet was identified as a resource to maximize efficiency in the process. It's quicker, more accurate, less costly, and hassle-free when compared to making phone calls and paging through hardbound directories. You can locate buyers at all levels and of all kinds, gaining access to phone numbers, e-mail addresses, submission policies, and more through a few simple clicks.

Whether you are selling a cheerleading instructional video, dramatic feature, or short documentary, search engines can be utilized to direct you to company Web sites that deal with product similar to your own. For example, by typing "independent sport instructional videos" in Google, site listings appear for cheerleading, kayaking, football, and baseball tapes. The Web sites featured are obviously good targets as possible distributors for like product. Whatever your specific subject matter or genre, a search engine can lead you to literally hundreds of companies that market films and videos similar to your own. Once at these sites, key contact information is only several clicks away.

VIRTUAL VIDEO STORES

Many potential Web-based buyers have no real "brick-and-mortar" existence, other than warehouses and shipping facilities. The majority of their business is conducted through the Internet with buyers that have landed on their site in search of product. Amazon is the biggie, and the company has a featured area strictly for independent product (though you can have your video or DVD listed in other, more appropriate areas, as well). You'll need a UPC to sell through this giant online retailer, otherwise you can only get product listed in the auction and market sections. Full instructions are given on the site detailing the process to submit programming, price and stock product, and when and how to expect payment.

Thousands of other video sellers exist on the Web in addition to *amazon.com*. From large electronic retailers like *buy.com*, *reel.com*, and *virginmega.com* to smaller players like *videoflicks.com* and *super-clubvideo.com*, from the very niche oriented sellers such as *horseonly.com* (which sells videos on proper tack techniques) to video rental store sites mirroring their land-based counterparts, it's not tough to find a seller appropriate for your film or video project. Each of these "stores" has its own policies and procedures for doing business with them; unfortunately, it's a different process in discovering this information with almost every site you visit. More often than not, there should be a "vendor" area or something that describes how to do business with the company. Like Ingram Entertainment (the large wholesale distributor), some sites offer exhaustive information on what an independent supplier should do to solicit the company. Others are not so formal, and you may find yourself electronically communicating with the president of the company (who also happens to be the Webmaster, packer, and shipper).

Regardless of the set-up, you should explore every possibility on the Web that represents a potential market or sale. It'll take some time to scour the endless number of product sellers. Try to do it in a systematic manner, starting with the biggest and working your way down the ladder. When surveying potential online resellers of your product, inquire as to:

- Format, length, and genre of projects acceptable
- Shipping and stock policies
- Sales volume for like product
- Payment methods
- Returns process
- Contracts
- Site traffic

Also investigate other vendors selling videos or DVDs through the site. These producers should be easy to find electronically. Just do a name or film/video title search on the Web and you'll likely end up at their personal site. Send them an e-mail asking about their experiences with the seller. You don't want to initiate business with a company that is going to disappear into cyberspace two months after you ship them a box of product. Question other producers about product movement, too. It's fine to have your film or video listed for sale on as many sites as possible. You don't want to waste too much time, however, setting up deals with online stores that do little-to-no business. There are just too many mega-sellers out there, and you want to associate with these businesses.

Though the markets and outlets for independent feature films, especially those dealing with the horror genre, are active and numerous, they pale in comparison to the opportunities available for instructional and special-interest segments. Videos and DVDs in this category have experienced a surge in sales thanks to the Internet. The prohibitive traditional advertising costs associated with marketing a unique product to its very targeted yet geographically diverse audience is no longer a hindrance to producers. Web sites heralding the activities, beliefs, and interests of almost every subject imaginable exist on the Internet and sell a full line of products (including videos and DVDs) through this technology. Specialty producers can reach an unlimited audience even for the most arcane subjects. But be sure to check each site's legitimacy before getting involved or shipping product.

INSIDER INDUSTRY INFORMATION

In addition to its sales, research, community, and promotional aspects, the Internet provides film and videomakers (nearly) free access to entertainment industry news and information once only accessible through pricey magazine subscriptions (*Variety* runs about $250 a year) and personal connections previously unavailable to most independent producers. Enlightening statistics, buyer data, executive insight, and sometimes even editorial "gossip" prepare those new to the industry by allowing for a better understanding of the trade. Spending a half-hour daily "scanning the trades" is a great way to gain an education on what is oft believed to be a closed-door business.

Chapter 27 Summary Points

- The Internet offers limitless promotional and sales opportunities to independent film and videomakers.

- A film or video's promotional Web site can be used to trigger many actions, including requests for further information, distribution offers, media contact, fan-building, and simple awareness.

- Send e-mail messages containing a Web-site link to potential buyers and interested media in an effort to drive valuable traffic to your site.

- One of the biggest cost-saving advantages to Web-based promotion is the ability to offer all or part of your project for a "screening" right on the user's desktop.

- Digital distribution is now a reality — online streaming sites exist that will offer your film or video to paying customers.

- Independent films and videos can be sold directly through your own Web site, much like any other online purchase operation.

- Electronic newsletters are a virtually cost-free method of keeping people aware of your project's happenings.

- Daily updating, membership privileges, electronic postcards, and Web-based awards are all good methods to maintain regular site traffic.

- Free Web advertising is available through search-engine listings, database submissions, Web-ring alliances, and link trading.

- Very targeted, low-cost paid advertising is plentiful in the form of banner ads and e-mail newsletter insertions.

- Join as many appropriate online communities — film/video industry or subject specific — as possible. Communicating with others via the Internet is one of the best ways to make valuable contacts.

CONCLUSION

The last 120,000 words or so will hopefully get you headed in the right direction toward attracting audiences, media, distribution, and sales to your independent film or video project. While the process of doing so is not an exact science, and one that increasingly depends on constantly changing technology, the bottom line of your success depends on two things: product and hustle. You gotta have something the market wants or needs and you gotta be willing to bust your hump to make that market take notice. With those two aspects firmly in place, it doesn't hurt to remember the following 10 Rules, too. Good luck!

The 10 Rules of Film Promotion and Self-Distribution

1. Don't quit your day job.
There will be no income from the activity of promoting and distributing your film or video for at least several months. So, unless you are independently wealthy, you'll need cash flow from other sources for personal needs.

2. Own a good cell phone.
Do you really want a potential distributor to talk to your three-year-old daughter? You also want to be able to work during what were once regular downtimes (work commutes, lunch breaks, etc.).

3. Get voice-mail, a pager, and e-mail.
You must utilize all forms of communication technology available to survive and succeed in the self-distribution business. Different time zones and instant access are two quick reasons to invest in these services.

4. Learn how to network.
Never let someone off the hook until they've given you the name of another who may help your cause. Be the biggest, loudest advertiser of your product.

5. Remember your film/video is a product.
You are not the only person in the world who has suffered the woes of creating an independent film or video. What you are offering, while precious to you, is not unique. There are always 20 people in line behind you with a similar project.

6. Keep good records.
Don't lose important contact names, phone numbers, and e-mail addresses. Devise a system to track calls, mailings, and sales in a systematic manner. Write down and keep everything.

7. Be tenacious.
You need to be unrelenting in your pursuit of publicity and sales, but...

8. Don't ask twice.
A "no" is a "no" forever in this business. You'll only appear unprofessional if you attempt to resolicit a potential buyer.

9. Ignore negativity.
Don't pay attention to the "no's," just the "yes's."

10. Only get involved with projects you are passionate about.
With all of the time, money, and effort involved in successfully promoting, distributing, and selling an independent feature film or video, this rule is obvious.

APPENDIX

FILM/VIDEO ATTORNEYS

John Cones
Marina City Drive, Suite 704W
Marina del Rey, CA 90292
Phone: (310) 477-6842
Email: *jcones@gte.net*

Mark Litwak, Esq.
Mark Litwak & Associates
433 N. Camden Drive, Ste. 1010
Beverly Hills, CA 90210
Phone: (310) 859-9595
Fax: (310) 859-0806
Email: *atty@marklitwak.com*

James N. Talbott
Talbott & Talbott
30765 Pacific Coast Highway, Suite 324
Malibu, CA 90265
Phone: (310) 457-1387
Fax: (310) 457-4425
Email: *jim@legalinterface.com*

Gregory Victoroff
Rohde & Victoroff
1880 Century Park East, Suite 411
Los Angeles, CA 90067
Phone: (310) 277-1482
Fax: (310) 277-1485

ENTERTAINMENT DISTRIBUTION COMPANIES

Artisan Entertainment
2700 Colorado Avenue
Santa Monica, CA 90404
Phone: (310) 255-3703

Avalanche Home Entertainment
2 Bloor St. W
Suite 1901
Toronto, ON, Canada   M4W 3E2
Phone: (416) 944-0104
Fax: (770) 442-9393

A Plus Entertainment
15030 Ventura Blvd.
Suite 762
Sherman Oaks, CA 91403
Phone: (818) 994-9831

Allied Entertainment Group
14930 Ventura Blvd.
Suite 304
Sherman Oaks, CA 91403
Phone: (818) 728-9900

Anchor Bay Entertainment
1699 Stutz
Troy, MI 48084
Phone: (800) 786-8777, ext. 4486

BFS Video
360 Newkirk Road
Richmond Hill
Ontario, Canada L4C 3G7
Phone: (905) 884-2323

BMG Entertainment
1540 Broadway
33rd Floor
New York, NY 10036
Phone: (212) 930-4000

Bruder Releasing Inc.
2020 Broadway
Santa Monica, CA 90404
Phone: (310) 829-2222, ext. 3

BWE
55 North 300 West
Suite 315
Salt Lake City, UT 84110
Phone: (801) 575-3680

Columbia TriStar Home Video
Sony Pictures Plaza
10202 W. Washington Blvd.
Culver City, CA 90232
Phone: (310) 244-7770

Crown International Pictures
8701 Wilshire Blvd.
Beverly Hills, CA 90211
Phone: (310) 657-6700

Creative Light Worldwide
8383 Wilshire Blvd.
Suite # 212
Beverly Hills, CA 90211
Phone: (323) 658-9166

Dead Alive
1933 West Main Street
Suite 5-200
Mesa, AZ 85201
Phone: (800) 214-6019

E.I. Independent Cinema
P.O. Box 132
Butler, N.J. 07405
Phone: (973) 283-2226

First Run Features
153 Waverly Place
New York, NY 10014
Phone: (212) 243-0600

Fox Searchlight
P.O. Box 900
Beverly Hills, CA 90212
Phone: (310) 369-4340

Full Moon Pictures
1645 N. Vine Street 9th Floor
Hollywood, CA 90028
Phone: (323) 468-0599

Goldhill Home Media
137 East Thousand Oaks Blvd.
Suite 207
Thousand Oaks, CA 91360
Phone: (805) 495-0735, ext. 207

Green Communications
1811 W. Magnolia Blvd.
Burbank, CA 91506
Phone: (818) 557-0050

Heartland Film Festival Video
200 S. Meridian, Suite 220
Indianapolis, IN 46225-1076
Phone: (317) 464-9405

Ideal Marketing
111 West Main Street
Mesa, AZ 85201
Phone: (480) 649-9688

IFM Film Associates
1328 East Palmer Avenue
Glendale, CA 91205
Phone: (818) 243-4976

Independent Pictures
42 Bond Street
6th Floor
New York, NY 10012
Phone: (212) 993-1235

J&M Entertainment
1289 Sunset Plaza Drive
Los Angeles, CA 90069
Phone: (310) 652-7733

Legacy Releasing
1800 Highland Avenue
Suite 311
Hollywood, CA 90028
Phone: (323) 461-3936

Leo Films
1701 N. Vermont Ave.
#108
Los Angeles, CA 90027
Phone: (323) 666-7140

Lion's Gate
561 Broadway
Suite 12B
New York, NY 10012
Phone: (212) 966-4670

Mainline
1801 Avenue of the Stars
Suite 1035
Los Angeles, CA 90067
Phone: (310) 286-1001

MGM/United Artists
2500 Broadway
Santa Monica, CA 90404
Phone: (310) 449-3000

Miklen Entertainment
1330 Schooner Lane
Anaheim, CA 92801
Phone: (714) 991-3751

MLR Films
59 Westminster Avenue
Bergenfield, NJ 07621
Phone: (201) 385-8139

Monarch Home Video
2 Ingram Blvd.
LaVergne, TN 37089
Phone: (615) 287-4000

MTI Video
14216 SW 136th Street
Miami, FL 33186
Phone: (305) 255-8684

New Line Home Video
116 N. Robertson
Los Angeles, CA 90048
Phone: (310) 854-5811
Fax: (310) 967-6678

Overseas Filmgroup/Firstlook Pictures
8800 Sunset Blvd.
Suite 302
Los Angeles, CA 90069
Phone: (310) 855-1199

Palm Pictures
4 Columbus Circle
5th Floor
New York, NY 10019
Phone: (212) 506-5800

Paramount Home Entertainment
5555 Melrose Avenue
Los Angeles, CA 90038
Phone: (323) 956-5000

Pathfinder Pictures
801 Oceanfront Walk
Suite 7
Venice, CA 90291
Phone: (800) 562-3330
www.pathfinderpictures.com

Raven Releasing
31 Cherry Street
2nd Floor
Millford, CT 06460
Phone: (203) 876-7630

RDF Los Angeles
225 Santa Monica Blvd.
Suite 1112
Santa Monica, CA 90401
Phone: (310) 301-6621

Redwood Communication
228 Main Street
Studio 17
Venice, CA 90291
Phone: (310) 458-6521

RGH/Lion's Share
8831 Sunset Blvd.
Suite 300
West Hollywood, CA 90069
Phone: (323) 823-5766

Samuel Goldwyn
9570 W. Pico Blvd.
Suite 400
Los Angeles, CA 90035
Phone: (310) 284-9220
Fax: (310) 860-3120

Santelmo
901 Wilshire Blvd.
#350
Santa Monica, CA 90401
Phone: (310) 656-0777

Seventh Art
7551 Sunset Blvd.
#104
Los Angeles, CA 90046
Phone: (323) 845-1455

Simitar Entertainment
5555 Pioneer Creek Drive
Maple Plain, MN 55359
Phone: (612) 479-7000

Sony Pictures Classics
550 Madison Ave.
8th Floor
New York, NY 10011
Phone: (212) 833-8500

Storm Entertainment
127 Broadway
Suite 200
Santa Monica, CA 90401
Phone: (310) 656-2500

Stratosphere Entertainment
767 5th Avenue
47th Floor
New York, NY 10153
Phone: (212) 605-1010

Summit Entertainment
1630 Stewart Street
Suite 120
Santa Monica, CA 90404
Phone: (310) 309-8400

Trimark
4553 Glencoe Ave.
Marina Del Ray, CA 90292
Phone: (310) 314-3096

Trident Releasing
8401 Melrose Place
2nd Floor
Los Angeles, CA 90069
Phone: (323) 655-8818

USA Films
9333 Wilshire Blvd.
Garden Level
Beverly Hills, CA 90210
Phone: (310) 385-6666
Fax: (310) 385-4400

Vanguard International Cinema
15061 Springdale St.
Suite 109
Huntington Beach, CA 92649
Phone: (714) 901-9020

Ventura Distribution
3543 Old Canejo Road
Suite 102
Newbury Park, CA 91320
Phone: (805) 498-5140

Very Video
1975 Stirling Road
Suite 201
Dania, FL 33004
Phone: (954) 921-8080, ext. 4

Vista Street Entertainment
9831 West Pico Blvd.
Los Angeles, CA 90035
Phone: (310) 556-3074

Warner Home Video
4000 Warner Blvd.
Bldg. 139, Room 25
Burbank, CA 91505
Phone: (818) 954-6978

Westar Entertainment, Inc.
10520 Venice Blvd.
Culver City, CA 90232
Phone: (310) 836-6795

Xenon Home Entertainment
1440 9th Street
Santa Monica, CA 90401
Phone: (310) 451-5510

York Pictures
16133 Ventura Blvd.
Suite 1140
Encino, CA 91436
Phone: (818) 623-9755

Zeitgeist Films
247 Centre Street
2nd Floor
New York, NY 10013
Phone: (212) 274-1989

INDEPENDENT-FRIENDLY FILM/VIDEO MARKETS AND EVENTS

American Film Market (AFM) - February
10850 Wilshire Boulevard
9th Floor
Los Angeles, CA 90024-4311
Phone: (310) 446.1000
Fax: (310) 446.1600
Email: afm@afma.com
www.afma.com/AFM/afm_home.asp

Independent Feature Film Market (IFFM) - late September
104 West 29th Street, 12th Floor
New York, NY 10001-5310
Phone: (212) 465-8200
Fax: (212) 465-8525
www.ifp.org/market24/

Videos Software Dealers Association Expo (VSDA) - July
Show Manager
Phone: (714) 513-8810
Fax: (714) 513-8848
*http://www.homeentertainment2002.com/videoshow/v31/index.cvn?
id=10000*

National Association of Broadcasters (NAB) - April
Phone: (202) 429-5300
Fax: (202) 429-4199
1771 N Street, NW
Washington, DC 20036
www.nab.org

IMPORTANT INDEPENDENT FILM/VIDEO PRINT MEDIA

Boxoffice Magazine
155 South El Molino Avenue
Suite 100
Pasadena, CA 91101
Phone: (626) 396-0250
www.boxoff.com

FILMMAKER
501 Fifth Avenue, Suite 1714
New York, NY 10017
Phone: (212) 983-3150
www.filmmakermagazine.com

Hollywood Reporter
5055 Wilshire Blvd.
Los Angeles, CA 90036-4396
Phone: (323) 525-2000
www.hollywoodreporter.com

The Independent
304 Hudson Street, 6th floor
New York, NY 10013
Phone: (212) 807-1400
www.aivf.org

Independent Feature Project Calendar
8750 Wilshire Blvd.
Beverly Hills, CA 90211
Phone: (818) 505-1942
www.ifp.org

MovieMaker Magazine
2265 Westwood Blvd., #479
Los Angeles, CA 90064
Phone: (310) 234-9234
www.moviemaker.com

Premiere Magazine
1633 Broadway
New York, NY 10019
Phone: (212) 767-5400
www.premiere.com

Variety/Daily Variety
5700 Wilshire Blvd., Suite 120
Los Angeles, CA 90036
Phone: (323) 965-4476
www.variety.com

Video Business
5700 Wilshire Blvd.
Los Angeles, CA 90036-5804
Phone: (323) 857-6600
www.videobusiness.com

Video Store Magazine
201 East Sandpointe Avenue, Suite 600
Santa Ana, CA 92707
Phone: (714) 513-8400
www.videostoremag.com

INDEPENDENT FILM/VIDEO PRODUCER REPRESENTATIVES

Jonathan Dana
1233 Ozeta Terrace
Los Angeles, CA 90069
Phone: (310) 273-0194

Jeff Dowd
225 Santa Monica Blvd.
Suite 610
Santa Monica, CA 90401
Phone: (310) 576-6655

Noel Lawrence
3450 Sacramento
#116
San Francisco, CA 94118
Phone: (415) 346-9682

Matthew Lesher
800 South Robertson Blvd.
Suite 8
Los Angeles, CA 90035
Phone: (310) 85-0071

Patrick Lynn
2617 S. Sepulveda
#9
Los Angeles, CA 90064
Phone: (310) 473-5237

Gary Meyers
2600 Kent Street
Berkeley, CA 94710
Phone: (510) 644-9131

A.I.P. Inc.
F. Joseph Clark
P.O. Box 3824
Citrus Heights, CA 95611
Phone: (916) 961-3784

AKA Movies, Inc.
Claude Kananack
6855 Santa Monica Blvd.
4th Floor
Los Angeles, CA 90038
Phone: (323) 769-7545

Berton and Donaldson
Mark Litwak
9595 Wilshire Blvd.
Suite 711
Beverly Hills, CA 90212
Phone: (310) 859-9595

Bertran & Donaldson
Michael Morales
9595 Wilshire Blvd.
Suite 711
Beverly Hills, CA 90212
Phone: (310) 271-5123, ext. 811

Film Kitchen
Doug Lindeman
7223 Beverly Blvd.
Suite 203
Los Angeles CA 90036
Phone: (213) 936-6677

Garvin, Davis & Benjamin
Tom Garvin
9200 Sunset Blvd.
Penthouse 25
Los Angeles, CA 90069
Phone: (310) 278-7300, ext. 23

Grainy Pictures, Inc.
John Pierson
75 Main Street
Cold Spring, NY 10516
Phone: (914) 265-2241

Harris Tulchin and Associates
Paul Hazen
Trident Center
11377 West Olympic
2nd Floor
Los Angeles, CA 90064
Phone: (310) 914-7979

Hollywood Broadcasting
Margo Romero
6363 Sunset Blvd.
4th Floor
Hollywood, CA 90028

Menemsha Entertainment
Neil Friedman, Paula Silver
1157 South Beverly Drive
2nd Floor
Los Angeles, CA 90035
Phone: (310) 712-3720

Metapictures
Sanford Rosenberg
2084 Union Street
San Francisco, CA 94123
Phone: (415) 563-2445

Redeemable Features
Melissa Chesman
381 Park Avenue
South Penthouse
New York, NY 10016
Phone: (212) 685-8585

Rudolph & Beer
Steven Beer
432 Park Avenue South
2nd Floor
New York, NY 10016
Phone: (212) 684-1001

Shooting Gallery
Ryan Werner
145 Avenue of the Americas
7th Floor
New York, NY 10013
Phone: (212) 243-3042, ext. 47

Sloss Special Projects
Micah Green
Joy Newhouse
170 5th Avenue
Suite 800
New York, NY 10010
Phone: (212) 627-9898

Sunshine Interactive Media
Jedidiah O. Alpert
Gregory Little
740 Broadway
New York, NY 10003
Phone: (212) 995-2222

Surpin, Mayersohn & Edelstone
Paul Mayersohn
1880 Century Park East
Suite 618
Los Angeles, CA 90067
Phone: (310) 553-6503

Swartz-Boyd
Michael Swartz
5604 Rhodes Avenue
Suite 105
North Hollywood, CA 91607
Phone: (818) 761-5886

The Movie Group
Peter Strauss
5750 Wilshire Blvd.
#5 THF
Los Angeles, CA 90036-3697

Tri Vision Entertainment
Hye Young Choi
4201 Wilshire Blvd.
Suite 518
Los Angeles, CA 90010
Phone: (323) 930-0167

Unicorn Films
Paul Tobias
37 Cranberry Street
Brooklyn, NY 11201
Phone: (718) 522-5870

United Talent Agency
Howard Cohen
9560 Wilshire Blvd.
5th Floor
Beverly Hills, CA 90212
Phone: (310) 273-6700, ext. 2071

WMA Independent
Rena Ronson
151 El Camino Drive
Beverly Hills, CA 90210
Phone: (310) 859-4315

HELPFUL INDEPENDENT FILM PROMOTIONAL WEB SITES

www.indieclub.com

www.filmmakers.com/films/directory.htm

www.indyfilm.buried.com/Independent_Films/

www.reeluniverse.com/Filmmaking/filmmaking.html

www.business2.com/webguide/0,1660,20961%7C382%7C0%7C0%7 C1%7Ca,00.html

http://dir.yahoo.com/Entertainment/Movies_and_Film/Titles/Independent/

www.moviefund

www.allfilmmarkets.com/index.htm

www.marklitwak.com/default.htm

www.indiewire.com

www.aivf.org

www.indierep.com

www.filmplaylinks.com

www.business.com/directory/media_and_entertainment/home_enter-tainment/distributors_and_wholesalers/

www.indiebin.com

www.tapelist.com

www.mwp.com

www.customflix.com

www.filmthreat.com

RECOMMENDED READING

43 Ways to Finance Your Feature Film. Cones, John. Carbondale, IL: Southern Illinois University Press, 1998

A Pound of Flesh. Linson, Art. New York: Grove Press, 1993

All I Need To Know About Filmmaking I Learned From The Toxic Avenger. Kaufman, Lloyd. New York: The Berkeley Publishing Group, 1998

The Computer Videomaker Handbook. Boston: Focal Press, 2001

Film and Video Financing. Wiese, Michael. Studio City, CA: Michael Wiese Productions, 1991

Film and Video Marketing. Wiese, Michael. Studio City, CA: Michael Wiese Productions, 1989

Film Finance and Distribution. Cones, John. Los Angeles: Silman James Press, 1992

Film Producing/Low-Budget Films that Sell. Harmon, Renee. Hollywood: Samuel French Trade, 1988

Filmmakers and Financing. Levison, Louise. London: Focal Press, 1994

Filmmaking on the Fringe. McDonagh, Maitland. New York: Carol Publishing Group, 1995

Filmmaker's Dictionary. Singleton, Ralph. Los Angeles: Lone Eagle Publishing, 2000

Hit and Run. Griffin, Nancy. New York: Simon & Schuster, 1996

How I Made a Hundred Movies in Hollywood and Never Lost a Dime. Corman, Roger. New York: Random House, 1990

How to Make Your Own Feature Movie for $10,000 or Less. Russo, John. New York: Barclay House, 1994

Independent Feature Film Production. Goodell, Gregory. New York: St. Martin's Press, 1982

The Independent Film and Videomaker's Guide. Wiese, Michael. Studio City, CA: Michael Wiese Productions, 1988

Killer Instinct. Hamsher, Jane. New York: Broadway Books, 1997

Making Independent Films. Stubbs, Liz. New York: Allworth Press, 2000

Making It In Film. London, Mel. New York: Simon & Schuster, 1985

Making Movies. Lumet, Sidney. New York: Random House, 1995

Media Marketing. Miller, Peter. New York: Harper & Row Publishers, 1987

Micro Budget Hollywood. Gaines, Philip and Rhodes, David. Los Angeles: Silman James Press, 1995

The Movie Business Book. Squire, Jason. New York: Fireside, 1988

Movie Marketing. Lukk, Tiiu. Los Angeles: Silman James Press, 1997

Ogilvy on Advertising. Ogilvy, David. New York: Random House, 1985

Opening the Doors to Hollywood. DeAbreu, Carlos. Los Angeles: Custos Morum Publishing, 1995

Persistence of Vision. Gaspard, John and Newton, Dale. Studio City, CA: Michael Wiese Productions, 1996

Producing, Financing and Distributing Film. Baumgarten, Paul. New York: Limelight Editions, 1995

Producing for Hollywood. Mason, Paul. New York: Allworth Press, 2000

The Publicity Handbook. Yale, David. Chicago: NTC Business Books, 1982

Rebel Without a Crew. Rodriguez, Robert. New York: Penguin Group, 1995

Selling Your Film. Sherman, Eric. Los Angeles: Acrobat Books, 1999

Slacker. Linklater, Richard. New York: St. Martin's Press, 1992

The Sleaze Merchants. McCarty, John. New York: St. Martin's Press, 1995

TV PR. Chambers, Wicke. Rocklin, CA: Prima Publishing, 1987

The Ultimate Film Festival Survival Guide. Gore, Chris. Los Angeles: Lone Eagle Publishing, 1999

What They Didn't Teach You at Film School. Landau, Camille and White, Tiare. New York: Hyperion, 2000

The Whole Picture. Walter, Richard. New York: The Penguin Group, 1997

Author Mark Steven Bosko has served in various capacities as writer, producer, director, and distributor on six independent features, including the cult classic horror comedy, *Killer Nerd*. That film was a widely popular video rental item in North America, and enjoyed sales success in such far-flung locales as Malaysia, Mexico, and Sweden. Made for $12,000, *Killer Nerd* could be found in major video chains across the country, was featured on CNN and *Entertainment Tonight*, and was eventually sold to Troma Pictures — where it became one of their most requested re-released films of all time.

Bosko's last produced film, *Pig*, was regionally released to the home video market, finding chain store sales success in New York, Florida, and Ohio. A cable premiere is in the works as the film continues to roll out to other states and is exploited via other markets. Bosko is also helming marketing activities for a new reality-based horror feature, *June Nine*, and is already fielding offers from several major and independent distributors expressing interest in representing the finished product in the home entertainment marketplace.

In addition to his direct filmmaking and distributing experience, Bosko works as a public relations and marketing professional, assisting the promotional campaigns for his own and other filmmakers' product through the Independent Film and Video Marketing Group (*www.IFVMG.com*), a company dedicated to helping independent filmmakers reach their promotional, distribution, and sales goals. Recent work includes Michael Wiese's upcoming feature, *Bali Brothers*, and the Web site *Itsonlyamovie.com*. You can e-mail Mark directly at *markbosko@IFVMG.com*.

FILM DIRECTING: SHOT BY SHOT
Visualizing from Concept to Screen

Steven D. Katz

Over 150,000 Sold! International best-seller!

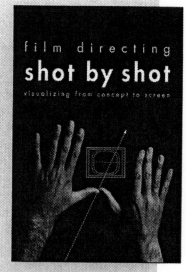

Film Directing: Shot by Shot — with its famous blue cover — is the best-known book on directing and a favorite of professional directors as an on-set quick reference guide.

This international bestseller is a complete catalog of visual techniques and their stylistic implications, enabling working filmmakers to expand their knowledge.

Contains in-depth information on shot composition, staging sequences, visualization tools, framing and composition techniques, camera movement, blocking tracking shots, script analysis, and much more.

Includes over 750 storyboards and illustrations, with never-before-published storyboards from Steven Spielberg's *Empire of the Sun*, Orson Welles' *Citizen Kane*, and Alfred Hitchcock's *The Birds*.

"(To become a director) you have to teach yourself what makes movies good and what makes them bad. John Singleton has been my mentor... he's the one who told me what movies to watch and to read *Shot by Shot*."
— Ice Cube, *New York Times*

"A generous number of photos and superb illustrations accompany each concept, many of the graphics being from Katz' own pen... *Film Directing: Shot by Shot* is a feast for the eyes."
— *Videomaker Magazine*

Steven D. Katz is also the author of *Film Directing: Cinematic Motion*.

$27.95, 366 pages
Order # 7RLS | ISBN: 0-941188-10-8

Making Movies?

Self-distribute them with CustomFlix!

Introducing CustomFlix — the new way to self-distribute your films worldwide. There's no inventory. No minimums. No exclusivity. And no hassles! A setup fee of only $49.95 gets you:

- ▶ Your own customizable e-store with streaming trailer for a whole year
- ▶ On-demand DVD duplication, as well as VHS tapes (we replicate too)
- ▶ Professional amaray-style packaging with full color printing and shrinkwrap
- ▶ Order fulfillment, monthly profit checks, 24/7 online reports, and more!

Our Online Promotion Program helps you sell on Amazon, Yahoo!, eBay, and Froogle, plus be visible on search engines such as Google and Alta Vista. We also author DVDs.

"Without CustomFlix, the Hoffman Collection would never have gotten off the ground. With their pay-as-you-go system, it made sense to release all my titles, not only the ones I was sure would move in volume. I heartily recommend the creative and tech-savvy folks at CustomFlix."
David Hoffman, Emmy Award-Winning Documentary Filmmaker

www.CustomFlix.com/mwp (888) 304-0049 Custom**Flix**

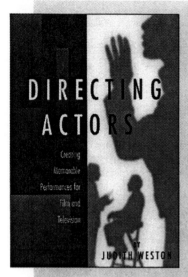

DIRECTING ACTORS
Creating Memorable Performances for Film & Television

Judith Weston

Over 20,000 Sold!

Directing film or television is a high-stakes occupation. It captures your full attention at every moment, calling on you to commit every resource and stretch yourself to the limit. It's the white-water rafting of entertainment jobs. But for many directors, the excitement they feel about a new project tightens into anxiety when it comes to working with actors.

This book provides a method for establishing creative, collaborative relationships with actors, getting the most out of rehearsals, troubleshooting poor performances, giving briefer directions, and much more. It addresses what actors want from a director, what directors do wrong, and constructively analyzes the director-actor relationship.

"Judith Weston is an extraordinarily gifted teacher."
— David Chase, Emmy Award-Winning Writer,
Director, and Producer
The Sopranos, Northern Exposure, I'll Fly Away

"I believe that working with Judith's ideas and principles has been the most useful time I've spent preparing for my work. I think that if Judith's book were mandatory reading for all directors, the quality of the director-actor process would be transformed, and better drama would result."
— John Patterson, Director
The Practice, Law and Order, Profiler

Judith Weston was a professional actor for twenty years and has taught Acting for Directors for over a decade.

$26.95, 314 pages
Order # 4RLS
ISBN: 0-941188-24-8

24 HOURS/ 1.800.833.5738   LOWEST PRICES AVAILABLE AT WWW.MWP.COM

SETTING UP YOUR SHOTS
Great Camera Moves Every Filmmaker Should Know

Jeremy Vineyard

Written in straightforward, non-technical language and laid out in a nonlinear format with self-contained chapters for quick, on-the-set reference, *Setting Up Your Shots* is like a Swiss army knife for filmmakers! Using examples from over 140 popular films, this book provides detailed descriptions of more than 100 camera setups, angles, and techniques — in an easy-to-use horizontal "wide-screen" format.

Setting Up Your Shots is an excellent primer for beginning filmmakers and students of film theory, as well as a handy guide for working filmmakers. If you are a director, a storyboard artist, or an animator, use this book. It is the culmination of hundreds of hours of research.

Contains 150 references to the great shots from your favorite films, including *2001: A Space Odyssey*, *Blue Velvet*, *The Matrix*, *The Usual Suspects*, and *Vertigo*.

"Perfect for any film enthusiast looking for the secrets behind creating film. Because of its simplicity of design and straightforward storyboards, *Setting Up Your Shots* is destined to be mandatory reading at film schools throughout the world."
— Ross Otterman, *Directed By Magazine*

Jeremy Vineyard is a director and screenwriter who moved to Los Angeles in 1997 to pursue a feature filmmaking career. He has several spec scripts in development.

$19.95, 132 pages
Order # 8RLS
ISBN: 0-941188-73-6

THE WRITER'S JOURNEY
2nd Edition
Mythic Structure for Writers

Christopher Vogler

Over 100,000 units sold!

See why this book has become an international bestseller and a true classic. *The Writer's Journey* explores the powerful relationship between mythology and storytelling in a clear, concise style that's made it required reading for movie executives, screenwriters, playwrights, scholars, and fans of pop culture all over the world.

Both fiction and nonfiction writers will discover a set of useful myth-inspired storytelling paradigms (i.e., "The Hero's Journey") and step-by-step guidelines to plot and character development. Based on the work of Joseph Campbell, *The Writer's Journey* is a must for all writers interested in further developing their craft.

The updated and revised second edition provides new insights and observations from Vogler's ongoing work on mythology's influence on stories, movies, and man himself.

"This book is like having the smartest person in the story meeting come home with you and whisper what to do in your ear as you write a screenplay. Insight for insight, step for step, Chris Vogler takes us through the process of connecting theme to story and making a script come alive."
> — Lynda Obst, Producer
> *Sleepless in Seattle, Contact, Someone Like You*
> Author, *Hello, He Lied*

Christopher Vogler, a top Hollywood story consultant and development executive, has worked on such high-grossing feature films as *The Lion King* and conducts writing workshops around the globe.

$24.95, 325 pages
Order #98RLS
ISBN: 0-941188-70-1

24 HOURS/ 1.800.833.5738   LOWEST PRICES AVAILABLE ONLINE AT WWW.MWP.COM

DIGITAL FILMMAKING 101
An Essential Guide to Producing Low-Budget Movies

Dale Newton and John Gaspard

The Butch Cassidy and the Sundance Kid of do-it-yourself filmmaking are back! Filmmakers Dale Newton and John Gaspard, co-authors of the classic how-to independent filmmaking manual *Persistence of Vision*, have written a new handbook for the digital age. *Digital Filmmaking 101* is your all-bases-covered guide to producing and shooting your own digital video films. It covers both technical and creative advice, from keys to writing a good script, to casting and location-securing, to lighting and low-budget visual effects. Also includes detailed information about how to shoot with digital cameras and how to use this new technology to your full advantage.

As indie veterans who have produced and directed successful independent films, Gaspard and Newton are masters at achieving high-quality results for amazingly low production costs. They'll show you how to turn financial constraints into your creative advantage — and how to get the maximum mileage out of your production budget. You'll be amazed at the ways you can save money —and even get some things for free — without sacrificing any of your final product's quality.

"These guys don't seem to have missed a thing when it comes to how to make a digital movie for peanuts. It's a helpful and funny guide for beginners and professionals alike."
> — Jonathan Demme
> Academy Award-Winning Director
> *Silence of the Lambs*

Dale Newton and John Gaspard, who hail from Minneapolis, Minnesota, have produced three ultra-low-budget, feature-length movies and have lived to tell the tales.

$24.95, 283 pages
Order # 17RLS | ISBN: 0-941188-33-7

ORDER FORM

TO ORDER THESE PRODUCTS, PLEASE CALL **24** HOURS **- 7** DAYS A WEEK
CREDIT CARD ORDERS **1-800-833-5738** OR FAX YOUR ORDER **(818) 986-3408**
OR MAIL THIS ORDER FORM TO:

MICHAEL WIESE PRODUCTIONS
11288 VENTURA BLVD., # 621
STUDIO CITY, CA 91604
E-MAIL: MWPSALES@MWP.COM
WEB SITE: WWW.MWP.COM

WRITE OR FAX FOR A FREE CATALOG

PLEASE SEND ME THE FOLLOWING BOOKS:

TITLE	ORDER NUMBER (#RLS _____)	AMOUNT
_____	_____	_____
_____	_____	_____
_____	_____	_____
_____	_____	_____
_____	_____	_____
	SHIPPING	_____
	CALIFORNIA TAX (8.00%)	_____
	TOTAL ENCLOSED	_____

SHIPPING:
ALL ORDERS MUST BE PREPAID, UPS GROUND SERVICE ONE ITEM **- $3.95**
EACH ADDITIONAL ITEM ADD **$2.00**
EXPRESS **- 3** BUSINESS DAYS ADD **$12.00** PER ORDER
OVERSEAS
SURFACE **- $15.00** EACH ITEM   AIRMAIL **- $30.00** EACH ITEM

PLEASE MAKE CHECK OR MONEY ORDER PAYABLE TO:

MICHAEL WIESE PRODUCTIONS

(CHECK ONE) ____ MASTERCARD ____VISA ____AMEX

CREDIT CARD NUMBER _____

EXPIRATION DATE _____

CARDHOLDER'S NAME _____

CARDHOLDER'S SIGNATURE _____

SHIP TO:

NAME _____

ADDRESS _____

CITY _____ STATE _____ ZIP _____

COUNTRY _____ TELEPHONE _____

ORDER ONLINE FOR THE LOWEST PRICES

24 HOURS | **1.800.833.5738** | **www.mwp.com**

Printed in the United States
75775LV00004B/64

9 780941 188760